ESSAYS ON CROP PLANT EVOLUTION

ESSAYS ON CROP PLANT EVOLUTION

EDITED BY

SIR JOSEPH HUTCHINSON

CAMBRIDGE
AT THE UNIVERSITY PRESS
1965

PUBLISHED BY
THE SYNDICS OF THE CAMBRIDGE UNIVERSITY PRESS

Bentley House, 200 Euston Road, London, N.W. 1
American Branch: 32 East 57th Street, New York 22, N.Y.
West African Office: P.O. Box 33, Ibadan, Nigeria

Printed in Great Britain by
Spottiswoode, Ballantyne and Co. Ltd
London and Colchester

CONTENTS

Foreword *page* vii

I The Beginnings of Agriculture in North West Europe
by H. GODWIN, *Botany School, University of Cambridge* 1

II The Evolution of Maize
by PAUL C. MANGELSDORF, *Harvard University, Cambridge, Mass., U.S.A.* 23

III The Development of the Cultivated Sorghums
by H. DOGGETT, *East African Agriculture and Forestry Research Organisation, Serere, Uganda* 50

IV The Comparative Phylogeny of the Temperate Cereals
by G. D. H. BELL, *Plant Breeding Institute, Cambridge* 70

V Cytogenetics and the Evolution of Wheat
by R. RILEY, *Plant Breeding Institute, Cambridge* 103

VI The History and Relationships of Cultivated Potatoes
by K. S. DODDS, *John Innes Institute, Bayfordbury, Hertford, Herts.* 119

VII The Evolution of Forage Grasses and Legumes
by J. P. COOPER, *Welsh Plant Breeding Station, Aberystwyth, Wales* 142

VIII Crop Plant Evolution: A General Discussion
by J. B. HUTCHINSON, *School of Agriculture, Cambridge, England* 166

Bibliography 182

Index 195

FOREWORD

In the study of evolution, the small group of plant species that have been domesticated, and have become crop plants, provide one of the most interesting and rewarding fields of enquiry. After de Candolle's classic study, interest waned and little further work was done until Vavilov and his Russian colleagues re-opened the subject after the first World War. Their wide-ranging collecting expeditions, their botanical studies on the collections they made, and Vavilov's penetrating analysis of the relation between origin and development on the one hand and variability on the other, set the stage for the numerous studies in crop plant evolution that have been reported in recent years.

In the Lent term of 1962 it was possible to arrange for a group of students of crop plant evolution and agricultural history to lecture in Cambridge on their own studies. The lectures provided an opportunity to compare the evolutionary history of a diverse group of crop plants against a background of the time scale of the development of agriculture in Western Europe. These lectures formed the basis of the essays that follow. They put together in one place the diverse ways in which the genotype of a crop plant may be built up under the influence of natural and human selection, and they provide an opportunity for comparative study of the significance of the major genetic phenomena in evolutionary change.

J. B. HUTCHINSON

SCHOOL OF AGRICULTURE
CAMBRIDGE

I

THE BEGINNINGS OF AGRICULTURE IN NORTH WEST EUROPE

by H. GODWIN

IN travelling across the English countryside, diversified as it is by pasture and ploughland, managed woodland and scattered copses, defined by the long lines of hedgerow and ditch, it is stimulating to recall in the mind's eye how that landscape must have appeared before our ancestors had broken the natural oak forest that formerly lay over it in a deep continuous mantle. It is difficult even for the ecologist to visualize that forest landscape, but without such an exercise of the imagination it will be difficult and misleading to imagine the true pattern of the nature and consequences of early agriculture.

Bearing this in mind, it will usefully introduce our subject to consider the actual techniques by which the botanist has obtained his information on how and when prehistoric agriculture first came to Western Europe.

One method of approach has been to identify those plant remains, such as seeds and fruits, that were associated with prehistoric settlements and that chance may have preserved through partial burning, burial or incorporation in water-logged deposits. This method has been followed with conspicuous success by Dr Hans Helbaek of the National Museum of Copenhagen, who has thus been able to reach back into the origins of cultivated plants not only in Europe but in the Middle East. We may recall that whilst in many instances he has made use of carbonized grains recovered from archaeological sites, he has also made very effective use of grain impressions left in clay potsherds. These impressions occur within cavities in the baked clay when firing consumed seeds or spikelets that had accidentally become embedded in the potter's material. From such evidence Helbaek has accumulated enough data to provide a general outline of the changing importance of different cereals throughout the successive prehistoric and early historic cultural stages, not only of

TABLE 1. *Plant remains identified in bog-burials of the Danish early Iron Age* (*Helbaek, 1954*)

+ = present, c = common, ab = abundant

	Borremose	Tollund	Grauballe
Ranunculus acris L. (Meadow Buttercup)	–	–	c
R. repens L. (Creeping Buttercup)	–	–	+
Fumaria officinalis L. (Common Fumitory)	–	–	+
Brassica campestris L. (Turnip, Naven)	+	(?)	–
Lepidium latifolium L. (Broad-leaved Pepperwort)	–	+	+
Thlaspi arvense L. (Field Penny-cress)	–	–	+
Capsella bursa-pastoris (L.) Moench. (Shepherd's Purse)	–	+	+
Erysimum cheiranthoides L. (Treacle Mustard)	–	+	+
Camelina linicola Sch. et Sp.	c	ab	+
Viola arvensis Murr. (Field Pansy)	–	c	+
Cerastium caespitosum Gilib. (Common Mouse-ear Chickweed)	–	–	+
Stellaria media L. (Chickweed)	–	+	+
S. graminea L. (Lesser Stitchwort)	–	–	+
Spergula arvensis L. (Corn Spurrey)	ab	ab	ab
Scleranthus annuus L. (Annual Knawel)	–	–	+
Chenopodium album L. (Fat Hen)	ab	ab	c
Chenopodium sp.	–	–	+
Linum usitatissimum L. (Cultivated Flax)	–	ab	c
Trifolium arvense L. (Hare's-foot)	–	–	+
Potentilla anserina L. (Silverweed)	–	–	(?)
P. argentea L. (Hoary Cinquefoil)	–	–	+
Aphanes arvensis L. (Parsley Piert)	–	–	+
Polygonum aviculare agg. (Knotgrass)	+	–	+
P. lapathifolium agg. (Pale Persicaria)	ab	ab	ab
P. persicaria L. (Persicaria)	–	–	c
P. convolvulus L. (Black Bindweed)	+	c	c
Rumex acetosella L. (Sheep's Sorrel)	ab	+	ab
R. acetosa L. Sorrel	–	(?)	–
R. crispus L. (Curled Dock)	–	–	+

Oxycoccus quadripetalus Gilib. (Cranberry)	+	–	–
Myosotis arvensis (L.) Hill. (Common Forget-me-not)	–	–	+
Solanum nigrum L. (Black Nightshade)	–	–	c
Veronica serpyllifolia L. (Thyme-leaved Speedwell)	–	–	+
Rhinanthus cf. minor L. (Yellow-rattle)	–	–	+
Brunella vulgaris L. (Self-heal)	–	–	c
Galeopsis tetrahit agg. (Common Hemp-nettle)	–	c	c
Plantago lanceolata L. (Ribwort)	–	+	ab
P. major L. (Great Plantain)	–	–	c
Achillea millefolium L. (Yarrow, Milfoil)	–	–	+
Matricaria inodora L. (Scentless Mayweed)	–	–	+
Lapsana communis L. (Nipplewort)	–	–	+
Leontodon autumnalis L. (Autumnal Hawkbit)	–	–	+
Sonchus asper (L.) Hill (Spiny Milk- or Sow-Thistle)	–	–	+
S. oleraceus L. (Milk- or Sow-Thistle)	+	–	–
Crepis capillaris (L.) Wallr. (Smooth Hawk's-beard)	–	–	+
C. tectorum L.	–	–	c
Juncus spp.	–	–	–
Luzula campestris (L.) D.C. (Sweep's Brush, Field Woodrush)	–	–	+
Heleocharis palustris (L.) R. Br.	(?)	–	–
Carex spp.	–	–	+
Phragmites communis Trin. (Reed)	–	–	+
Seiglingia decumbens (L.) Bernh. (Heath Grass)	–	–	+
Lolium perenne L. (Rye-grass, Ray-grass)	(?)	–	ab
L. remotum Schr.	–	–	(?)
Poa nemoralis L. (Wood Poa)	–	–	+
Poa sp.	+	–	+
Bromus sp.	–	–	ab
Agropyron caninum R. et S. (Bearded Couch-grass)	–	–	c
Hordeum tetrastichum	–	ab	–
Avena fatua L. (Wild Oat)	–	(?)	+
Holcus lanatus L. (Yorkshire Fog)	(?)	–	ab
Deschampsia caespitosa (L.) Beauv. (Tufted hair-grass)	–	–	c
Phleum spp.	–	–	+
Echinochloa crus-gali (L.) Beauv. (Cockspur)	–	+	c
Setaria viridis (L.) Beauv. (Green Bristle-grass)	–	–	+

Britain, but also of large parts of Western Europe. We can only indicate the nature of his very far-reaching results by mentioning one or two examples. His data are sufficient, for instance, to illustrate the pronounced prevalence of wheat and barley cultivation in the British Neolithic and Bronze Ages and the relatively late (Roman) introduction and expansion of oats and rye in Britain (Helbaek, 1952). Similar evidence collected by Jessen and Helbaek (1944) indicates that flax growing was already

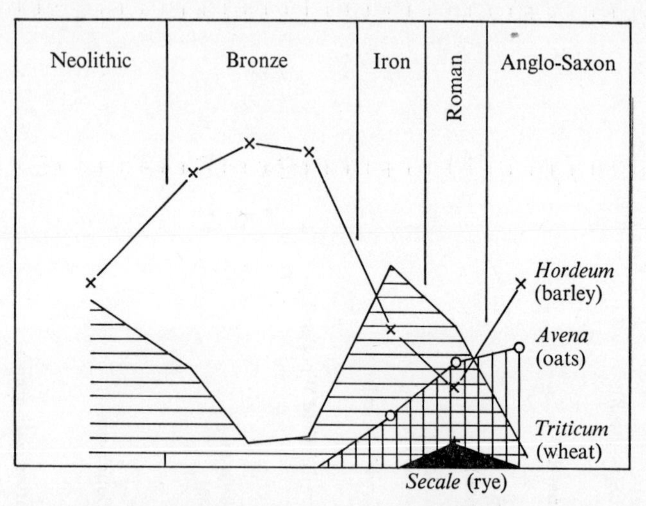

Fig. 1. The diagram is based upon data by Jessen and Helbaek (1944) and shows for each archaeological period the proportion of all British sites in which cereal grains or grain impressions have been recognized. (From Godwin, 1956.)

practised in England, Scotland and Ireland during the Bronze Age and elsewhere in Europe also by Neolithic people. Helbaek has been able to show that the wheat principally associated with early Neolithic husbandry from Mesopotamia and Egypt to Western Europe was Emmer, i.e. *Triticum dicoccum*, and at Jarmo he has certainly come close to the identification of the earliest cultivated cereals.

Not always, however, is the plant evidence preserved in either of the two forms mentioned. Thus there have been recovered on various occasions from Danish peat bogs, the well preserved corpses of people presumably buried in some ritual manner. Most notable of these was the so-called 'Tollund Man', whose stomach contents were so remarkably well preserved that Helbaek (1950) was able to reconstruct from them a good estimate of the man's last meal before he was strangled and interred in the bog. Table 1 illustrates the range of plant remains identified in three different bog burials of the same period and seems to indicate a gruel of the kind reported as the staple food of the tribes with whom the Romans fought in the West European lowlands. The range of weed plants that are found alongside the barley and flax suggests the possibility that the Danish Iron Age people collected seeds from plant communities resembling a weedy fallow, but the results also make us aware of the possibility that some of the plants now regarded as weeds may at that time have been deliberately cultivated. Indeed, in a later publication, Helbaek (1960*a*) makes a good case that *Chenopodium album*, the Goosefoot or Fat Hen, was actually thus cultivated, and discoveries from the remains of Iron Age settlements in Denmark have established the point that the seeds of individual 'weeds' were separately harvested, separately stored and no doubt had their own separate uses. In all three bog burials, the stomachs contained the achenes of *Polygonum lapathifolium* in such large amount that they must almost certainly have been deliberately collected.

The notion that plants no longer employed for food may have played an important part in prehistoric agriculture is further illustrated by the discovery made in a Late Bronze Age ditch filling on Rockley Down, Wiltshire, where along with a considerable amount of barley were found the abundant charred tubers of the onion couch grass, *Arrhenatherum tuberosum*. This is not known nowadays to be a food plant, but it seems difficult to suppose that it occurred by chance in this situation with few 'weed' associates, and it seems likely that it was an economically useful if not an actually cultivated plant.

The other important method of approach to analysis of the origins of agriculture consists in the study of the progressive changes in the natural vegetation cover of the countryside, a study most readily approached, and perhaps only approachable by, pollen analysis. The constant sedimentation into growing deposits of all kinds of a rain of airborne pollen grains and plant spores, to the extent of hundreds per square centimetre per annum, leads to the growth of sedimentary columns exceedingly rich in these sub-fossil indices of former vegetation. At any given level the contained pollen grains, when concentrated, examined and identified microscopically, reflect a picture of the vegetation from which they were derived. Likewise, the changes in pollen frequency from sample to sample throughout a deposit indicate the progress of vegetational changes whilst the deposit was being formed.

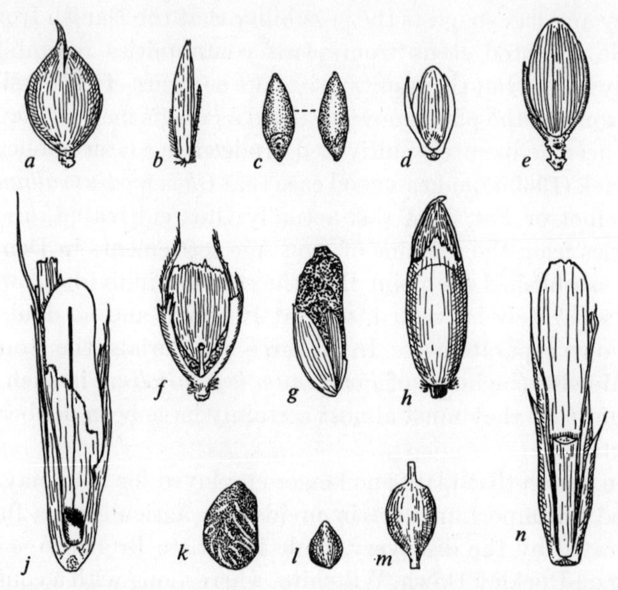

Fig. 2. Fruits and seeds of wild grasses and other plants recovered from the stomach of the Iron Age Grauballe man, a Danish bog burial. *a*, *e*, floret of *Echinochloa crus-galli*; *b*, floret of *Poa nemoralis*; *c*, grain of *Deschampsia caespitosa*; *d*, floret of *Setaria viridis*; *f*, *h*, floret of *Sieglingia decumbens*; *g*, floret of *Holcus lanatus* containing an ergot; *j*, floret of *Agropyron caninum*; *k*, achene of *Potentilla erecta*; *l*, fruit of *Alchemilla*; *m*, fruit of *Carex*; *n*, floret of *Lolium perenne*. (After Helbaek, 1958.)

Over large parts of Western Europe the accumulation of organic sediments recommenced at the opening of the Post-glacial period about 8000 B.C. and it has continued to the present day. Very large numbers of complete pollen diagrams for deposits of the Post-glacial period permit the reconstruction of past vegetational history throughout North-Western Europe. They reveal a pattern of vegetational history of remarkable consistency, which in general terms was that of the first spread of closed birch forest over open tundra or taiga, the replacement of birch forest by pine, and then the spread of mixed-oak-forest with hazel leading to a dominance of alder-mixed-oak-forest, which persisted until beech and hornbeam began to expand in the last two or three millennia before the present. The congruity of the diagrams indeed permitted the establishment of a system of pollen zonation for North-Western Europe, with a quasi chronological basis now in process of confirmation and adjustment by radiocarbon dating. It is not seriously doubted that this sequence of changes in vegetational dominance was primarily caused by the amelioration of climate that accompanied the close of the last Ice Age and the disappearance of ice sheets from Western Europe: indeed, the critical date of 8300 B.C. that marks the beginning of the Post-glacial period is the time at which the ice is known to have commenced its retreat from the last major line of moraines that crosses central Norway, Sweden and Finland.

Within the period of time covered by the pollen zones III to VIIa/VIIb, there is very little sign of human interference with the natural vegetation of Western Europe. Up to this time prehistoric man was in the Mesolithic culture stage, i.e. that of the hunter, fisher and food collector, subsisting upon the products of the natural environment, inevitably living in small nomadic groups, and without permanent settlements. He was one component of the ecosystem, dependent upon the forest, rivers and lakes for food, surrounded and dominated by the forest. A type site for such people was excavated at Star Carr near Seamer in East Yorkshire, by Professor Grahame Clark and members of the Cambridge Department of Archaeology and Sub-department of Quaternary Research (Clark *et al.*, 1954). There proved to have been a settlement of red-deer hunters established on the sandy margins of a large lake, within the deep organic muds and peats of which were incorporated and preserved the organic as well as the mineral remains of the occupation. An abundant microlithic flint and bone industry was recovered and it was shown to come from the boundary between pollen zones IV and V, that is to say, from the time when the hazel was beginning to spread swiftly into a landscape previously dominated by birch with some associated pine. The radiocarbon date for the Mesolithic platform found in the reeds of the lake margin was about 7500 B.C. Although the platform had been constructed of felled birch trees and artefacts were very numerous, the effects of the occupation upon the local vegetation were evident only in the infrequent pollen of a few nitrophilous plants such as docks, nettles and members of the goosefoot family that presumably reacted to the increased

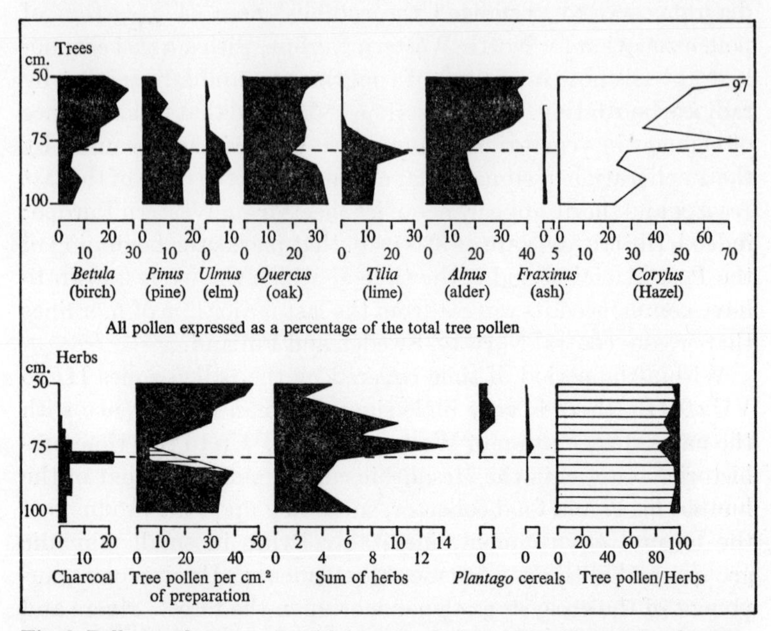

Fig. 3. Pollen analyses at a Danish bog site, Ordrup Mose, illustrating the effect of a local Neolithic forest clearance upon the local vegetation. In the lower of the two diagrams note the abrupt increase in the frequency of charcoal in the pollen preparations, the abrupt decline in absolute tree pollen frequency and the introduction of pollen of cereals and of plantain.
The upper diagram shows the decline of elm, oak, lime and pine at the clearance level, and the swift recovery afterwards of woody plants that are typically pioneers of recolonization, namely birch, hazel and alder. The failure of elm and lime to recover is striking. (After Iversen, 1941.)

organic content of a site where butchering and cooking were taking place.

If the modification of the natural closed cover of mixed-oak-forest by Mesolithic man was slight, the case was different with the advent of the first peoples to practise plant and animal husbandry. Our first realization of the nature and extent of the effects they produced came with the publication in 1941 of an historic paper 'Land occupation in Denmark's Stone Age' by

Dr J. Iversen (1941) of the Danish Geological Survey. He reported that in many Danish pollen diagrams there occurred changes of pollen frequency indicative of remarkably sudden alteration in the composition and condition of the forests. Thus at Ordrup Mose, a site where there was known to have been a Neolithic settlement beside the margin of a small lake, the pollen diagram showed a sudden fall in the frequency of pollen of the main mixed-oak-forest dominants (oak, elm and lime) shortly followed by substantial rises in the pollen frequency of hazel, birch and alder. These changes, which Iversen gave reasons for thinking were probably not climatically caused, were accompanied by a fall in absolute pollen frequency and by abundant remains of small charcoal stratified into the lake deposits. There was therefore good *a priori* reason to suspect that the changes might be attributed to clearance by the Neolithic occupants of the settlement. This conjecture was strongly justified by accompanying increases in the frequency of the pollen of grasses, *Artemisia* (mugwort) and Chenopodiaceae, all plants in the category of ruderals or weeds. The counts at the same level included, moreover, grass pollen of large diameter that was tentatively identified as that of cereal. Most important however, was the recognition of the earliest appearance of the pollen of the ribwort plantain, *Plantago lanceolata*, which is readily identified and distinct from the pollen of other species of plantain: since this plant is characteristic of open herbaceous vegetation and is quite intolerant of woodland shade, the inference that forest clearance had taken place must be extremely strong. It seemed almost certain that there had been local clearance of the natural mixed-oak-forest with planting of cereals, and that after abandonment, the cleared area had been re-invaded by birch, alder and hazel, all plants that are characteristic early recolonists of cleared woodland. The diagrams indicated, in this instance, a complete recovery of the original forest cover.

Iversen's paper showed that similar phenomena could be recognized in a number of Danish Neolithic settlement sites and he interpreted the observed effects as those of a *Brandwirtschaft* in which the primeval forest was cut, burned, and subjected to temporary cereal growing, possibly in association with grazing of cattle on the coppice shoots of the felled trees and shrubs.

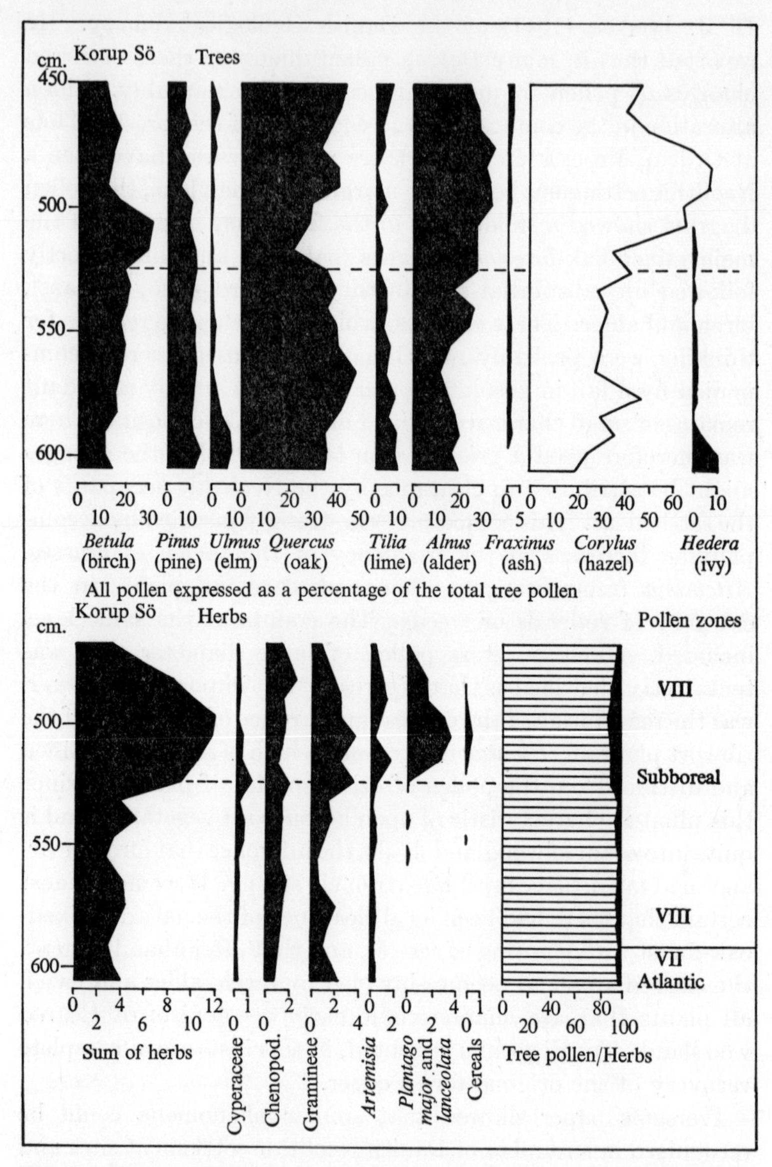

Fig. 4. Pollen analyses at a Danish lake site, Korup Sö, showing the effects of Neolithic forest clearance at the level of the broken horizontal line. The upper figure shows the decline of the mixed-oak-forest (especially oak and lime) and the consequent increase of birch and hazel. The lower diagram shows the great increase in frequency of herbaceous plants, including plantains and wormwood, along with cereals. (After Iversen, 1941.)

No sooner had Iversen published his findings and conclusions than it was apparent that his results coincided closely with our own from the Northern part of the East Anglian Breckland, that great expanse of largely treeless heaths, given over since medieval times to sheep grazing and rabbit warrens. In the long pollen diagrams from Hockham Mere in the Northern Breckland, the lake deposits could be seen to register the same pronounced fall in elm pollen frequency as occurred in the Danish diagrams, and with a corresponding introduction of the pollen of *Plantago lanceolata* and of grasses and heaths. From this level in the deposits the representation of herbaceous plant communities increased right up to the time when the lake was drained in late medieval times. There was already good reason to equate the sudden decline in elm pollen frequency with the opening of the Neolithic period in Western Europe, and the Breckland itself was known to be a region of outstandingly dense Neolithic settlement: the great Neolithic flint mines of Grimes Graves were only a few miles distant from the site of the pollen analyses at Hockham Mere. The probability thus began to emerge that it was not the dry climate or the poor soils which in themselves were responsible for the general treelessness of the Breckland, but that it was primarily due to agricultural clearances that had begun in the early Neolithic and had increased in extent thereafter (Godwin, 1944).

In the early 1950's Dr Iversen and his colleague, J. Troels-Smith, decided to make the crucial experiment of determining whether natural woodland could indeed be cleared by the means available to Neolithic man. They worked in an area of residual mixed-oak-forest at Draved in Northern Jutland and demonstrated that by the use of polished Neolithic axes (borrowed from the National Museum) it was quite feasible to clear woodland at an economic rate. They secured the advice of experts familiar with surviving fire clearance cultures and burned over the cut timber within the clearing, sowing the primitive cereals *Triticum monococcum* and *T. dicoccum* in the cooled ash. They succeeded in producing good crops in this manner and it was interesting to observe how quickly the weeds of cultivation accompanied them.

Since the time of the first publication of Iversen's paper the elm decline and associated signs of agricultural activity have been recognized all over Western Europe. The effect is so widespread that there has been natural speculation as to whether the populations of Neolithic man were large enough to have produced it, or whether it might not alternatively have been caused directly by climatic change. Everywhere, however, it seemed to correspond with the beginning of the Neolithic culture, so that very great interest has been associated with the attempt to find an absolute date for it in a large number of sites. This is a task to which radiocarbon dating has been successfully applied and we now have large number of dates, all of them suggesting an age of round about 3000 B.C. This is considerably older than the age that archaeologists had guessed for the beginning of the Neolithic in Western Europe, although not older than palynologists had thought likely for the corresponding pollen zone boundary. A considerable further number of radiocarbon dates upon charcoal from Neolithic settlements, and from settlements at the transition from Mesolithic to Neolithic, have supported this early date.

One such investigation is that made at Shippea Hill, Cambridgeshire, where the early Neolithic artefacts were stratified into the deposits of the abandoned channel of the river Little Ouse. Separate age determinations of the charcoal of the Neolithic occupation yielded dates of 2910 $\pm$ 120 and 2990 $\pm$ 120 B.C., whilst the peat just above and below the Neolithic culture layer gave ages (for no presently understood reason) a few hundred years older. The elm decline at this site, although present, was not very pronounced, but this was the level at which the pollen curve for *Plantago lanceolata* first appeared and the level at which bones of sheep were also found (see Fig. 5).

The West European date of 3000 B.C. fits well into the growing assembly of radiocarbon dates for the spread of Neolithic culture from its centre of origin in the Middle East, that is to say, Northern Iran, Irak, Turkistan, Palestine, Syria and the Caspian Shore. In Jericho it appears that about 6000 B.C. there flourished a proto-Neolithic culture with querns, domestic animals, and settlements with defences that themselves implied a non-nomadic existence and some measure of husbandry and food storage. In Jarmo, Northern Irak, a settlement with emmer, spelt and two-rowed barley, besides domestic animals, has been dated to about 4500 B.C. In Macedonia, Dr Willis has just obtained a radiocarbon date for the early Neolithic of 6000 B.C., and other early Neolithic dates are found in the Balkans. The early Neolithic *Bandkeramic* people spread northwards through Central Europe occupying the loose fertile soils of the Loess and after forest clearance growing emmer, einkorn, barley, peas, beans and flax: they reached Holland by 4000 B.C. and it is significant that the radiocarbon dates there and in South Germany were obtained directly by combustion of the prehistoric cereals themselves. More recently radiocarbon dates

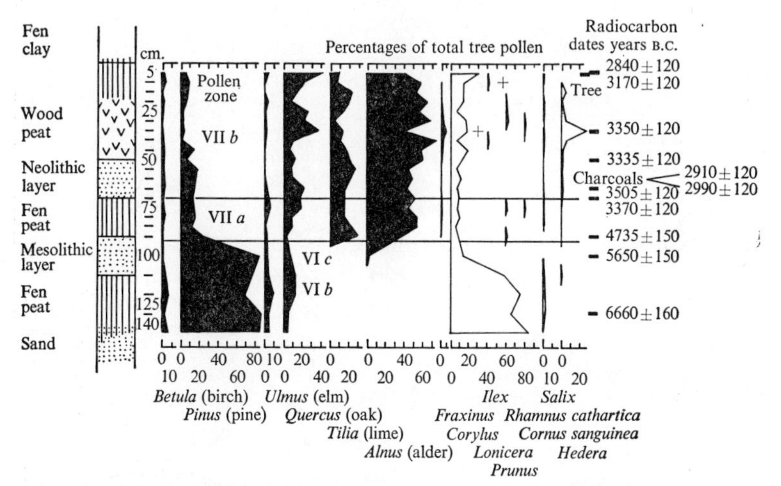

Fig. 5. Tree pollen diagram from the excavated site at Shippea Hill in the East Anglian fenland. Radiocarbon dates are shown from peat at different depths in the old river channel, including the sandy black layer containing Neolithic artefacts. Radiocarbon dates for charcoal in this layer are also shown. The tree pollen curves show little sign of forest clearance. (Clark and Godwin, 1962.)

have been published for early Neolithic settlements in Western France, Ireland and South-Western Sweden, between 3500 and 3000 B.C., so that spread must have been early along the Atlantic coast line (Fig. 6).

The reality and the date of the earliest agricultural clearance being established, the next step must be to clarify its nature and consequences. Dr Troels-Smith, working at Aamosen and employing very large and precise pollen counts, was able to demonstrate that the elm decline itself preceded, by a small but clear amount, any expansion of such weeds as those in the genera *Plantago*, *Rumex* and *Artemisia*. He therefore suggested that the earliest clearance was associated with a first stage of animal husbandry, with tethered stock fed entirely by leafy shoots of deciduous trees. He suggested that the elm was selectively gathered for this purpose, so causing the decline of elm pollen frequencies *before* the stage of burning and cereal cultivation described by Iversen. The practice of gathering tree foliage for fodder has in fact persisted up to the present day in many parts

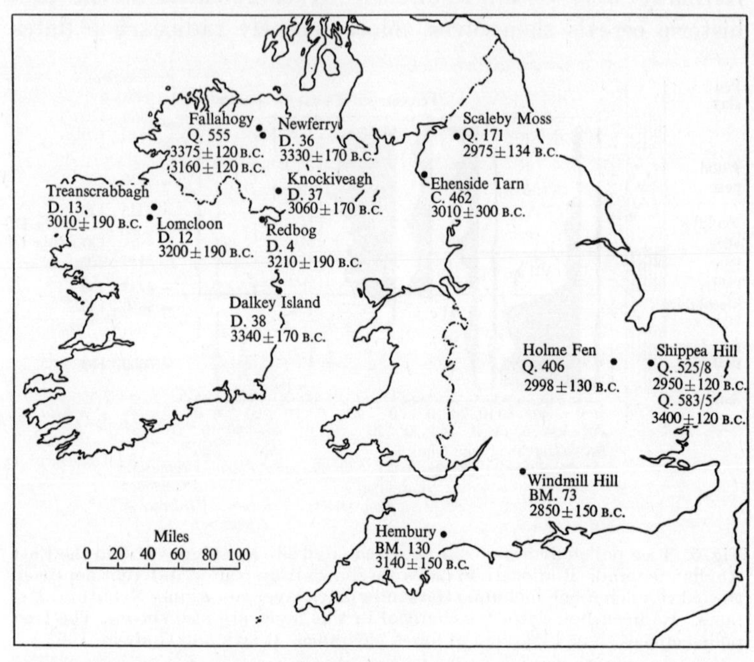

Fig. 6. Map showing those early Neolithic sites in the British Isles for which radiocarbon dates are available. (Clark and Godwin, 1962.)

of the world and has been responsible for the curiously devastated aspect of great areas of Himalayan and other forests. Moreover the elm is a tree much used in primitive cultures for this and related purposes. Subsequently, in investigations of Swiss Neolithic lake settlements, Troels-Smith found evidence of dung layers containing abundant leafy shoots and apparently associated with animals tethered in stalls. Investigators in Britain have also sometimes detected a time gap between the elm decline and the first appearance of indicators of open pasture, although the evidence is not altogether clear as to the meaning of this in vegetational terms.

We must of course not expect prehistoric land utilization at a time so remote as the early Neolithic to have existed in a shape familiar today, and between modern intensive agriculture and the earliest shifting cultivation within the deciduous forest there must have been many variants in the kind and intensity of exploitation and many transient systems of treating the modified vegetation types that arose successivly with the intake of different soils and different landscapes. Sometimes, none the less, there is fairly clear evidence of some particular stage or manner of land usage. This is the case at a site where a prehistoric wooden trackway has been uncovered at Blakeway Farm in the

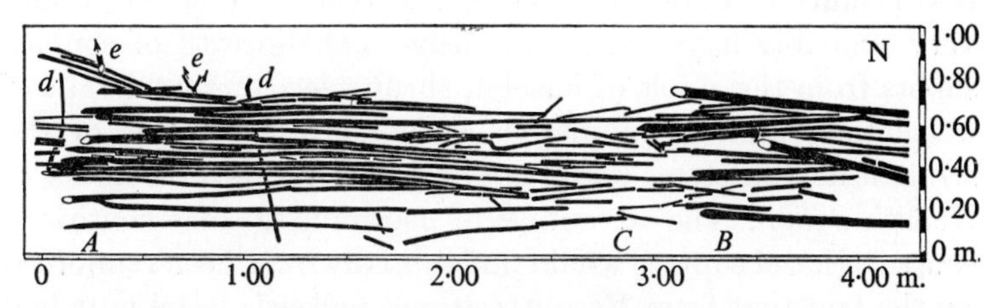

Fig. 7. Scale plan of a Neolithic trackway from the Somerset Levels. Nearly a mile long, it was constructed of faggots of hazel poles laid longitudinally: these were so straight and unbranched that they must have grown in conditions resembling coppice wood. Radiocarbon date, 2500 ± 130 B.C. (Godwin, 1960.)

Somerset Levels. This path, set upon the contemporary bog surface, consists of a narrow corduroy of successive faggots of long straight rods of hazel (*Corylus*) laid longitudinally side by side upon a thin layer of transversely placed bushes of heather. Here and there the poles lay upon transverse rods and the whole was secured by slender oblique pegs. The Cambridge radiocarbon date for the wood of the track was approximately 2500 B.C., that is to say fairly early in the Neolithic period. Not only were the hazel rods exceptionally straight and unbranched, but their length and their growth rings showed them to have been grown very rapidly and uniformly. There can therefore have been little doubt that they were grown in conditions closely approximating to those of hazel coppice of the present day. No doubt they grew in the woodlands upon the Wedmore Ridge from which the track runs southwards for about a mile to the island of Westhay. If we consider pollen diagrams from this region and embracing the same period of time, we shall find that the early Neolithic was marked firstly by a great decline in elm pollen, and secondly by great increases in the frequency of *Corylus* pollen. Since the sites of these pollen analyses generally lie in big peat bogs a mile or more distant from the nearest hills where the hazel was growing, we have to suppose that there must have been extremely widespread hazel coppice at this time. One is put in mind of the hazel scrub that grows at the present day over the Burren of Western Ireland, from which limestone country the trees themselves have long ago been removed. Startling as it is at first glance to visualize extensive coppice growth as early as the first part of the Neolithic period, we must recognize that primitive forest clearance, such as Iversen described, would very probably have led fortuitously to the growth of coppice shoots from the stools of hazel, a shrub exceedingly tolerant of mutilation, and the numerous economic applications of the straight flexible stems would not long have escaped the notice of Neolithic man. The advantage of maintaining some continued source of hazel coppice would undoubtedly have been reinforced by the fact that from Mesolithic times onwards hazel nuts had been gathered and used as a concentrated form of food highly suitable both to storage and transport. We may note in passing that the value of hazel coppice to an agricultural system is indicated by its persistence as a standing system of management right through medieval times up to the present day.

From the time of the earliest Neolithic forest clearances we may expect that although in some instances the openings were allowed to revert to woodland, in many others they persisted as coppice or woodland under-grazed by stock and in others again converted to rough pasture. These clearances would be extended from time to time by fresh intakes from the woodland according to local or general population pressures. From sites in many parts of the country and at very different periods the pollen diagrams show repeated minor episodes of forest clearance. Thus at Shapwick Heath, Miss J. Turner (1962) has recently shown that at a level carbon dated about 2000 B.C., and therefore late Neolithic, (and long after the elm decline with the earliest forest clearances) there was a forest intake affecting both the elm and the lime with associated increases in the pollen of herbaceous plants, bracken and the ash, together with only small increases in the pollen frequency of plantain and grass. Likewise at Whixall Moss, Shropshire (Fig. 8) Miss Turner showed that a similar decline in the lime was associated with increases in grass, plantain, bracken and miscellaneous herbs. It appeared that some form of grassland had been created by the clearance and although it was more at the expense of *Tilia* than of the other trees, it was evidently not due to selective felling of the one species, but seemed rather to be due to the preferential clearance of particular areas of ground in which the lime tree was growing abundantly. The likelihood that prehistoric people in fact would clear areas of specially favourable soil whilst leaving others, is strong. M. E. S. Morrison in Northern Ireland has cited a case from the Neolithic in which a limestone patch within a general area of basalt seems to have been singled out for early clearance by Neolithic people. Miss Turner's site at Whixall Moss is carbon dated at about 1280 B.C., so that it must be attributed to the Bronze Age. This site has the great advantage that carbon datings at different levels in the peat give one a measure of the speed of the vegetational processes indicated by the pollen analyses. As the samples are in fact only separated from one another by an interval of thirty years the time taken for the vegetational changes corresponds well with ecological expectations, the main effects expressing themselves in about 150 years.

In the late Bronze Age and early Iron Age the Urnfield people from Central Europe introduced the ox-drawn scratch plough, reconstructions of which have recently been made and tested in Reading University. The length of furrow achieved by the single ox was short and cross ploughing produced the characteristically small and square Celtic fields, which in all probability represent the first permanent arable land. The population increase in Central Europe at this time was considerable and the Urnfield people rapidly moved out to the western fringe of the Continent where their activities are faithfully reflected in the pollen diagrams by sharp increases in the indications of prehistoric agriculture. Thus in Somerset, at the time when the numerous Late Bronze Age trackways came to be constructed across the peat bogs of the Levels, there is a very big rise in the frequency of all

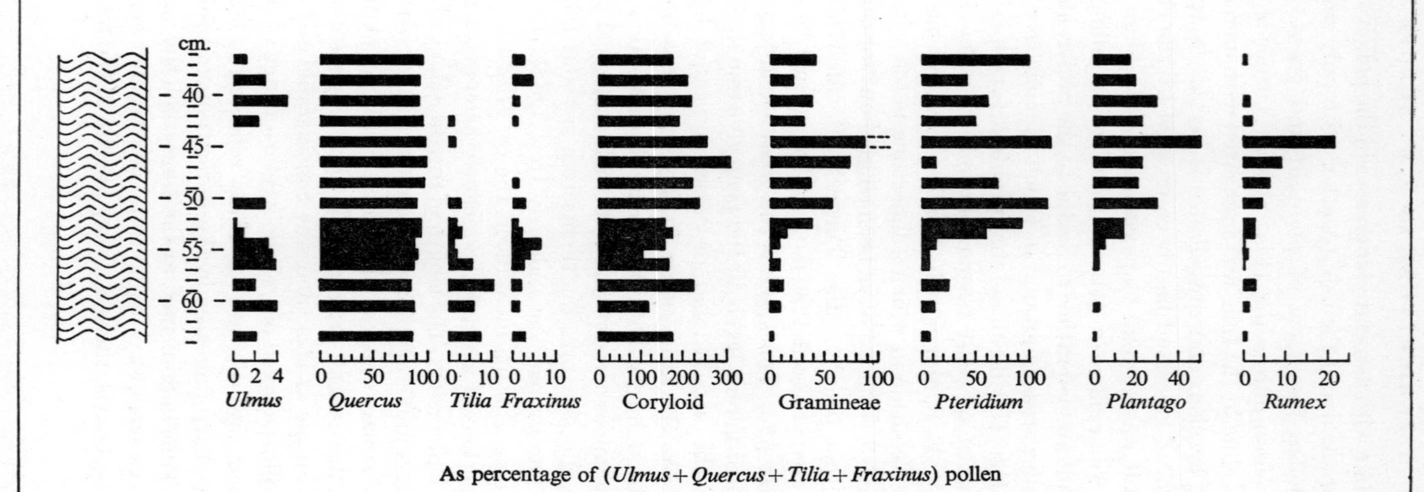

Fig. 8. Closely spaced pollen analyses made at Whixall Moss, Shropshire to elucidate the changes at the level of the *Tilia* (lime) decline. Each sample (1 cm thick) took about thirty years to form and the radiocarbon age determination for the episode is 1280 ± 115 B.C. Shortly after the *Tilia* decline, *Fraxinus* (ash) briefly expands, and then both this and *Ulmus* (elm) decrease greatly. The decrease is clearly associated with great expansion of grasses, bracken (*Pteridium*), plantains (*Plantago*) and docks (*Rumex*). The inference is that the cleared area became a form of grassland. (Miss J. Turner.)

the indicators of agriculture (Fig. 9) and similarly at Holme Fen, Huntingdonshire, on the edge of the East Anglian fenland where, after the first slight indications of Neolithic clearance at 3000 B.C., there is a large rise at levels carbon dated to about 1450 B.C., that is to say about the middle of the Bronze Age. Both in Somerset and Huntingdonshire the agricultural indicators

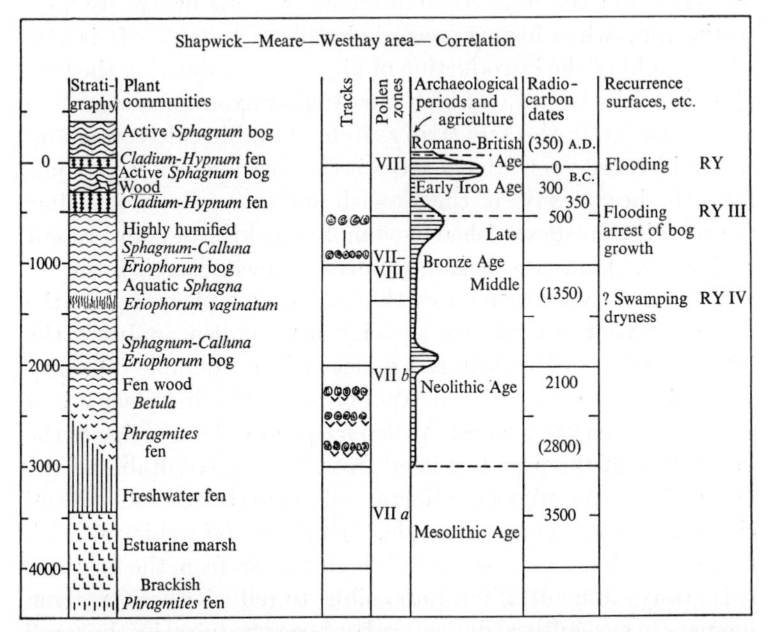

Fig. 9. Correlation table of events and horizons represented in the peat bogs of the Somerset Levels. The intensity of prehistoric agriculture is represented by the shaded curve showing the frequency of pollen of weeds, grasses and similar indicators. Low frequencies prevail in the Neolithic period from about 3000 B.C., there is a substantial maximum in the Late Bronze Age, and a much larger one in the pre-Roman Early Iron Age during the time of occupation of the lake-villages of Glastonbury and Meare.

include such a great variety of herbaceous types that they must certainly represent a considerable degree of arable cultivation. They include cereals and other grasses, plantains (including *P. lanceolata* as well as other species), chenopods, docks, nettles, umbellifers, Caryophyllaceae and Compositae (including wormwood, cornflower and many other types). In both sites also, the late Bronze Age agricultural activity was succeeded, after a period of quiescence, by a still more active period of agricultural activity. In Somerset this can be fairly confidently associated with the pre-Roman Iron Age, for this was the time of the building and the occupation of the well known lake villages at Glastonbury and Meare. Likewise the site at Holme Fen is very close to the big Roman station of Caistor by Peterborough and there is extremely abundant archaeological evidence for intensive Roman occupation of the East Anglian fenlands. This brings us close to the time when historic records become available. It is also within sight of the introduction of big ox teams drawing the true plough, with the aid of which our Anglo-Saxon ancestors were able to maintain the acre strip system of arable cultivation and with which they were able to exploit the clearances of woodlands from the heavy clays of the English lowlands. From this time onward the forests vanished even more quickly, and our present agricultural landscape began to emerge in recognizable form.

It must of course be realized that in all attempts to deduce the nature of past agricultural activity from pollen analyses, the interpretation of the data is a matter of considerable difficulty and calls for much ecological experience. In the first place it is of course necessary to separate pollen frequency changes due to the vegetation of the peat bog itself from those attributable to the vegetation of the adjacent mineral soil where the settlements and clearances were actually made. This can be achieved with experience, but a more serious difficulty arises from the fact that it is always difficult, if not impossible, to tell whether the given increase in agricultural indicators has been produced by the small clearance near to the pollen-sampling site, or whether it reflects more widespread change due to numerous settlements spread over a wider area. On the resolution of this basic difficulty our progress largely depends.

In some instances we are fortunate enough to recover macroscopic plant remains that reinforce conclusions based upon pollen analysis. This is so at two sites on the chalk downs of Kent, an area of particular botanical interest, not only because south-eastern England displays such climatic affinity with the continental mainland, but because it has been exposed, more than any other part of Britain to successive waves of invaders, and the impact of fresh cultures. We have in fact hitherto had little good evidence of when the Downs assumed their grassland

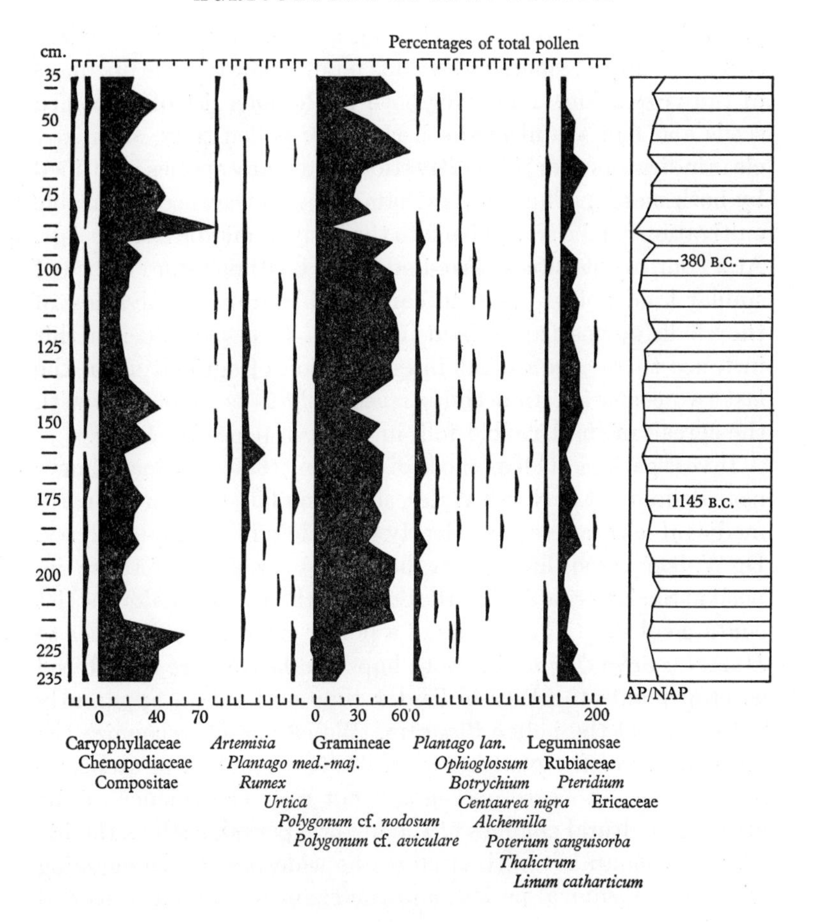

Fig. 10. Pollen diagram from a peat deposit at Wingham, at the eastern end of the Kentish Downs. The low tree pollen frequencies and the great variety of pollen of plants of open habitats and weeds of crops, indicates that the region was substantially under agriculture from at least 1600 B.C., as judged by the radiocarbon dates at the levels shown.

character. Now at Wingham in Kent a deep peat has yielded both pollen and macroscopic plant remains that bear directly on this problem (Fig. 10). Radiocarbon dates suggest that the deposit embraces a period from about 1600 B.C. to the Roman occupation. The low ratio of tree to herb pollen indicates a general condition of disforestation, and the very wide variety and frequency of small herbaceous plants indicates the openness of the vegetation. Moreover both in the long list of identified seeds and fruits, and in the herbaceous pollen there are many clear indicators of arable cultivation, and many species identified by both means. Thus at Wingham at least clearance and arable cultivation can be traced back to the early or middle Bronze Age. At a comparable site at Frogholt, near Eastbourne, evidence of similar kind points again to considerable arable cultivation of the chalk downs from about 1200 B.C. to *c.* 500 B.C.: in this instance there was a great intensification of agriculture in the last two centuries, an activity reasonably to be associated with the spread of the Urnfield folk in the area (Godwin, 1962).

Investigations such as those of Seddon (1960) at Dinas Emrys near Beddgelert in North Wales, show that late Roman and early medieval clearances are clearly recorded in pollen analyses. Dr Walker (1955) has shown that the upper layers of a lake deposit at Skelsmergh Tarn in the South of the English Lake district contain substantial amounts of a pollen grain referred either to *Humulus* or to *Cannabis*. Both hops and hemp were introduced as crop plants together in Tudor time and this is apparently witnessed by the pollen diagram. We can easily recognize the latest phase of the planting of exotic trees, and there seems no reason why with care one should not recover evidence of the great agricultural changes of the historic period, such as the big change from arable cultivation to the widespread sheep grazing of the late medieval period, and the changes that accompanied the abandonment of villages and such widespread catastrophes as the Black Death.

II

THE EVOLUTION OF MAIZE

by PAUL C. MANGELSDORF

MODERN maize, the starting point in a study of the evolution of the maize plant, is in most respects a typical grass but is unique among the major cereal grasses in the nature of its inflorescences (Weatherwax, 1955). The terminal inflorescence, commonly called the 'tassel' (Fig. 11, *a*, *c*), usually bears only male spikelets. These occur in pairs, one member of each pair being sessile the other pedicellate. Each spikelet contains two florets, each floret has three pollen sacs or anthers which at maturity are packed tightly with some twenty-five hundred pollen grains. These are small, about $\frac{1}{250}$ of an inch in diameter, light in weight, and are easily carried by the wind.

The lateral inflorescences (Fig. 11, *a*, *b*), which as they mature become the familiar 'ears', have only female spikelets. Like the spikelets of the tassel, these occur in pairs; both are sessile but one is potentially pedicellate, and may become so under certain circumstances. Each spikelet contains two florets, one, the upper, functional and the other, the lower, aborted or rudimentary. Each floret contains a single ovary, terminated by a long style, the pollen-receptive organ commonly known as the 'silk'. The silks are covered with fine hairs and are admirably designed to capture wind-blown pollen (Fig. 11, *e*). In contrast to the majority of cereals, maize is a naturally cross-pollinated plant and it is this feature, more than any other, that has made possible the production of hybrid maize, one of the most important and far-reaching developments in applied biology of this century (Mangelsdorf, 1951).

Each silk represents a potential kernel and must be pollinated for its ovary to produce a kernel. The kernels themselves are firmly attached to a rigid axis, the 'cob', and are not completely covered as are those of other cereals by the floral bracts which botanists call 'glumes' and which the layman knows as 'chaff'.

Instead the entire ear is enclosed, often quite tightly, by modified leaf sheaths, the husks or shucks (Fig. 11, *b*). Thus, while in other cereals the kernels are protected individually, in maize they are covered *en masse*. The result is that cultivated maize has no mechanism for the dispersal of its seeds and hence is no longer capable of reproducing itself without man's intervention. The

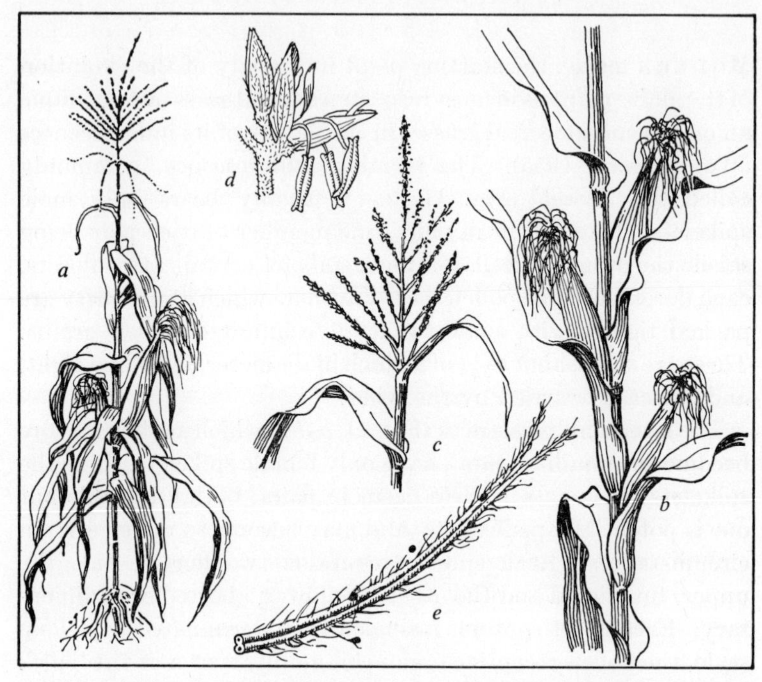

Fig. 11. Botanical characteristics of the modern corn plant. *a*. The entire plant showing the male inflorescence, the tassel, at the tip of the stalk and the female inflorescences, the ears, in the middle region. *b*. Young ears enclosed in husks with the pollen-receptive organs, the silks, protruding from the ends. *c*. Typical tassel. *d*. Typical male flower with three anthers containing pollen. *e*. A single silk magnified to show hairs and adhering pollen grains. (Drawing by G. W. Dillon.)

very characteristics which make maize so useful to man render it incapable of existing in nature. It is probable that maize would soon become extinct if deprived of man's protection.

How, then, did the wild ancestor of maize differ from cultivated maize in ways which enabled it to exist in nature for thousands or perhaps millions of years before man appeared on the scene? If wild maize had ears enclosed in husks, how did it disperse its seeds? If it did not have ears enclosed in husks, how were its seeds protected from the depredations of birds and insects? These are questions which we hoped to answer in part by the examination of the fossil and archaeological remains of prehistoric maize and in part by genetic and cytological studies of living maize and its relatives.

The fossil evidence of maize comprises a number of pollen grains isolated from a drill core taken from a depth of more than 200 ft below the present site of Mexico City in preparation for the construction of Mexico's first skyscraper. These were recognized as pollen grains of a grass—though unusually large ones—by Paul Sears of Yale University and Katherine Clisby of Oberlin College, who, employing pollen analysis to detect climatic changes, were engaged in studies of the drill core. The pollen was identified by Elso Barghoorn (1954) of Harvard University as that of maize which has the largest pollen of any known grass. In making his identification and in distinguishing between the pollen of maize and its two relatives, teosinte and *Tripsacum*, Barghoorn relied not on size alone but upon the ratio of the axis length to the diameter of the pore. The use of the axis/pore ratio to differentiate between pollen of maize and its relatives has recently been challenged by Kurz *et al.* (1960) who, measuring pollen of a variety of maize grown under sixteen different experimental conditions, found both the axis length and pore diameter to be affected by the environment. These authors did not, however, deny that the largest of the fossil pollen grains were those of maize, and subsequent studies by Barghoorn* and his colleagues under phase microscopy have verified the earlier identifications and shown that some of the fossil pollen is unquestionably that of maize.

Although at least eighty thousand years old, since it is assigned to the last interglacial period, and probably the pollen of wild maize since man had not yet reached this part of the world, the fossil pollen is scarcely distinguishable in size, shape, and other characteristics from modern maize pollen (Plate I A). This fact permits two important conclusions to be drawn: (1) the ancestor of maize was maize and not one of its two American relatives, teosinte or *Tripsacum*; (2) maize is an American plant and did

* Personal Communication.

not have its origin in Asia as some botanists have recently suggested, (Stonor and Anderson, 1949) or in any of those other parts of the Old World considered by earlier students of plants to have been its ancestral home (cf. Mangelsdorf and Oliver, 1951; Mangelsdorf and Reeves, 1959*b*).

Further proof, if more is needed, of the American origin of maize is furnished by archaeological remains. In the Old World not a single prehistoric specimen of any part of the maize plant has ever been found, while in both North and South America parts of the maize plant including cobs, entire ears, husks, and tassels, are among the most common of the vegetal remains found in archaeological sites.

The oldest known remains of maize come from a once-inhabited rock shelter, called Coxcatlán Cave, after the village in the southern part of the state of Puebla, Mexico, near which it is located. This cave has been excavated by Dr Richard MacNeish (1961, 1962) of the National Museum of Canada. Cobs of a tiny-eared maize were found in the next to the lowest level in the cave which has since been dated by radiocarbon determinations of other vegetal remains at 5000 B.C. The fact that these cobs are quite uniform in size and other characteristics suggests that they may represent wild maize or maize in the earliest stages of domestication.

A detailed botanical study of the cobs from Coxcatlán Cave has not yet been completed.* The oldest thoroughly-studied remains of maize come from another once-inhabited rock shelter, in New Mexico, known as Bat Cave. This was excavated by Herbert Dick, of the Peabody Museum of Harvard University and later of the Colorado University Museum, in two expeditions in 1948 (Mangelsdorf and Smith, 1949) and 1950. The cave was inhabited for several thousand years by people who practised a primitive form of agriculture and an even more primitive pattern of sanitation. During the centuries of their occupancy, garbage, excrement, and other debris accumulated in the cave to a depth of six feet creating exactly the kind of site into which archaeologists delight to dig. At the bottom of this accumulation

* A description by Mangelsdorf *et al.* of the prehistoric wild maize uncovered in Coxcatlán Cave and several other caves in the Valley of Tehuacán has recently appeared (*Science*, 1964, Vol. 143: 538).

of trash, Dick turned up some tiny cobs of ancient maize which have been dated by radiocarbon determinations of associated charcoal at about 5600 years (Plate I B).

The photograph reproduced in Plate II A shows one of the earliest Bat Cave cobs compared with two modern ears of maize, the Corn-Belt maize of the United States and the large-seeded flour maize of the Urubamba Valley of Peru. This picture poses the principal question to which virtually all of our research has in recent years been addressed: 'How did this little Bat Cave maize, or something like it, evolve into these highly developed races of North and South America even in 5600 years?'

To answer this question we had first to determine the botanical nature of the Bat Cave maize and to identify those of its characteristics which were clearly primitive. My associate, Dr W. C. Galinat, and I made an intensive study of one of the Bat Cave specimens which contained the partial remains of a single kernel. Each part of this cob was carefully dissected under the microscope and measured. On the basis of the measurements, Galinat prepared the diagramatical, longitudinal section illustrated in Fig. 12. The tiny kernels which this cob must once have borne could only be those of popcorn, a type in which the kernels are small and hard and capable of exploding when exposed to heat. The stalks or pedicels on which the kernels were borne, much longer than those of modern maize, and the long foliaceous floral bracts which almost completely enclosed them, show that the Bat Cave maize was also a form of pod corn, a type in which the individual kernels are enclosed in glumes or chaff (Plate IIB).

Pod corn has been considered by many botanists, beginning with Saint-Hilaire (1829), to be a primitive ancestral form. Sturtevant (1894), a long-time student of maize, concluded many years ago that both popcorn and pod corn are primitive. My former colleague, R. G. Reeves, and I (Mangelsdorf and Reeves, 1939), on the basis of quite different evidence, reached a similar conclusion. The ancient Bat Cave specimens provide convincing archaeological evidence in support of these conclusions.

The discovery that the Bat Cave maize is a form of pod corn imparted renewed interest to the studies, in which we had for some years been engaged, on the genetic and morphological nature of pod corn. This peculiar type, which still occurs as a 'rogue' or 'freak' in some South American varieties, is in some localities preserved by the Indians who believe it to have magical properties (Cutler, 1944). Pod corn has also sometimes been grown in gardens in the United States as a curiosity. Today it is most likely to be found in the experimental cultures of maize geneticists who maintain it for its 'marker' gene, designated as *Tu*, on the fourth longest chromosome of maize.

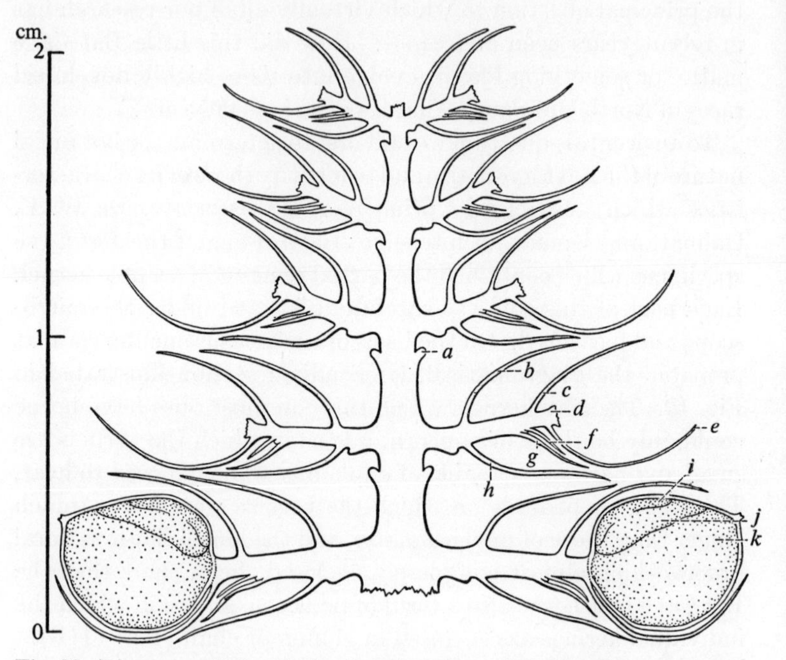

Fig. 12. Diagrammatic longitudinal section of one of the Bat Cave cobs based on measurements of dissected parts. The tiny kernels show that this was a pop-corn; the long pedicels on which the kernels are borne and the bracts which almost enclose them indicate that it was also a pod corn. (Drawing by W. C. Galinat.)

One result of our intensive study of pod corn has been the discovery of an intermediate allele, designated tu^h, at the *Tu-tu* locus. By incorporating, through repeated backcrossing, the two higher alleles in the series, *Tu* and tu^h, into a uniform inbred strain, originally of the genotype *tutu*, it has been possible to produce six distinct genotypes as follows: *TuTu*, $Tutu^h$, *Tutu*, tu^htu^h, tu^htu, and *tutu*. A comparison of a large number of characteristics of these six genotypes, which are isogenic except for the *Tu-tu* locus and such genes as are closely linked to it, has now been made and the data have been briefly summarized in a preliminary statement (Mangelsdorf and Mangelsdorf, 1957) and will be treated in full detail elsewhere. The data which are pertinent to the present discussion are considered below.

The data show that the *Tu-tu* locus is strongly pleiotropic affecting many different characteristics of the maize plant. In proceeding through the series of the six genotypes from *TuTu* to *tutu* the following profound changes occur: (1) a reduction in the terminal inflorescence, the tassel, accompanied by an increase—but a substantially greater one—in the lateral inflorescence, the ear; (2) a change from a predominantly pistillate tassel to a wholly staminate one; (3) a reduction in the number of tassel branches accompanied by an increase in the prominence of the central spike; (4) a progressive decrease in the length and the weight of the glumes accompanied by an increase in the size and weight of the rachis. All of these changes—if they occurred during the evolution of maize under domestication, as we now suppose that they did—tended to make the maize plant less able to survive in the wild and more useful to man.

The photograph in Plate II A, which depicts a typical ear of Corn Belt dent corn of the United States, illustrates the end product of the several changes initiated by mutations at this single locus on chromosome 4: (1) the ear is borne on a short, thick branch near the middle position of the stalk; (2) it is wholly pistillate and grain bearing; (3) like the central spike of the tassel it is unbranched; (4) in the absence of enveloping glumes the larger part of the available energy goes into rachis and grain. An ear of this size may weigh as much as three-fourths of a pound. It would be difficult to imagine an ear of this weight borne by any of the other cereals, rice, wheat, barley, or even sorghum.

There have, of course, been many other loci involved in the evolution of maize under domestication. The changes at the *Tu-tu* locus have themselves been magnified by the development of modifier complexes until the point has been reached where the homozygous pod corn of the genotype *TuTu* is, on most modern genetic backgrounds, monstrous and sterile. And there have been evolutionary forces other than gene mutation contributing to the evolution of maize. Yet it is this single locus more than any other which set the plant on new evolutionary paths which, in a man-made environment, led to its becoming one of the world's most successful cultivated species. In its far-reaching effects and in its importance to mankind, this locus is perhaps comparable to that on chromosome 5 of wheat, described by Riley in this volume, which resulted in the diploidization of the chromosomes of the polyploid wheats. Indeed it may be that one of the principle differences between evolution in nature and evolution under domestication is that mutations with large effects, which are likely to be lost in nature, are preserved under domestication. In maize and in other cultivated species there have been a number of mutations which have come close to meeting the specifications of Goldschmidt's 'hopeful monster' —mutations which affect the entire organism so profoundly that a new species, or something resembling one, is created almost literally overnight.

There is no doubt that pod corn is primitive in a number of its characteristics: in having its caryopses enclosed in glumes as do other cereals; in bearing both pistillate and staminate spikelets in the tassel; in providing a means of dispersal for its seeds. Despite this fact and because it is often monstrous and sterile, it has been dismissed by a number of botanists from any rôle in the ancestry of maize (cf. Mangelsdorf and Reeves, 1959*a*). We believe that its monstrousness has been misunderstood, that pod corn is monstrous today because it is a 'wild' relict character superimposed upon modern highly domesticated varieties. Today's pod corn is comparable to a primitive 1900 chassis powered by the sophisticated engine of the latest model car. The surprising thing is not that pod corn is sometimes monstrous but that it is not more so—that the locus governing its expression is capable of functioning at all in a milieu so different from that in which it had evolved and to which it was undoubtedly once well adapted. We have assumed that pod corn would be less monstrous and would exhibit normal grass characteristics when combined with other 'wild' genes and we hoped to find these in varieties of popcorn, a type which we regarded as primitive; a conclusion which the Bat Cave maize seemed to support.

Our hopes have been realized. A number of varieties of popcorn tend to modify the characteristics of pod corn when crossed with it and some do so quite drastically. The varieties Lady Finger and Argentine carry complexes of modifying genes which appreciably reduce the monstrousness of pod corn and a third variety, Baby Golden, carries a major modifying gene, designated as *Ti*, on chromosome 6 which acts as an inhibitor of the tunicate gene reducing its expression by approximately half.

By combining these modifying and inhibiting genes from these three popcorn varieties with the pod-corn gene we have developed a number of strains of popcorn which are homozygous for the *Tu* locus and which breed true for the pod-corn character. Some of these homozygous strains are much less monstrous than the usual forms of pod corn, are completely fertile and might under suitable conditions be capable of surviving in the wild.

The majority of these true-breeding pod-popcorns have other characteristics which we now regard as primitive. The plants when grown on fertile soils have several stalks instead of one as do most modern maize varieties and in this respect resemble the majority of wild grasses, including all of the known relatives of maize both American and Asiatic. The plants are shorter than those of ordinary maize because one of the numerous effects of the pod-corn gene is to shorten and thicken the upper internodes of the stalk. This shortening causes, or at least is accompanied by, the development of a terminal inflorescence which bears both staminate and pistillate spikelets, the former above, the latter below on the same tassel branches. These branches breaking apart easily when disturbed by the wind or by birds, provide one of the most important primitive characteristics which modern cultivated maize lacks, a mechanism for the dispersal of its seeds.

Plants of homozygous pod-corn frequently do not have ears —most of their energy is apparently concentrated in the terminal inflorescences—but when they do have ears these are borne high upon the stalk often at the joint of the stem immediately below the tassel. This elevation of the position of the ear has a number of important consequences which are illustrated in Fig. 13. The diagram, which is based on data from several many-eared plants, shows how a number of the characteristics of the ears are determined by their position on the stalk: (1) the higher the position, the smaller the ear, partly for the simple mechanical reason that the stalk at this position is slender and is incapable of bearing a heavy load; (2) the higher the ear the more likely it is to have both staminate and pistillate spikelets; (3) the higher the ear the shorter the lateral branch or 'shank' upon which it is borne; the shorter the branch the fewer the joints from which the husks arise; the fewer the husks the less completely the ear is enclosed. Thus an ear borne immediately below the tassel is enclosed while the young seeds are developing but as these mature the husks flare open allowing the ear to disperse its seeds. In brief a simple change in position determined by a single-gene change can provide a mechanism for dispersal of the seeds borne on the ear as well as those borne on the fragile branches of the tassel.

These facts seem to answer several of the most puzzling questions involved in previous attempts to explain the evolution of maize: How could wild maize have survived the handicap of an ear incapable of dispersing its seeds and if wild maize had no ears how could the ear of modern maize, its most important organ, have come into existence?

The most primitive ear we have so far obtained by combining popcorn and pod corn is shown in Plate II C in comparison with an ear of modern dent maize (Plate II A) and with the most primitive cob dated at 4445 ± 180 years from La Perra Cave in Mexico which was excavated by Richard MacNeish.

In weight and number of kernels our reconstruction is much closer to the prehistoric specimen than to the ear of modern dent maize. The modern ear weighs 317 g, the ear of pod-popcorn weighs 1·99 g. However, only twenty-four of its thirty-eight female spikelets developed kernels. Had all done so it would weigh 2·47 g assuming the additional kernels to have the same average weight, 0·034 g, as those which are present. The La Perra specimen weighs only 0·52 g but it lacks both the forty-eight kernels which it once bore and a staminate spike. Without its kernels and its staminate spike, the reconstructed ancestral form weighs 0·87 g only slightly more than the prehistoric specimen.

Although we have not yet completely reconstructed wild maize or duplicated exactly the most primitive specimens from either Bat Cave or La Perra Cave—the glumes of the pod-popcorn are still too prominent to match those of the prehistoric specimens—we have succeeded in developing what is probably the world's most unproductive maize. This is useful in suggesting that we are on the right track in attempting to retrace maize's evolutionary paths.

The reconstructed ear illustrated in Plate II C has pistillate spikelets on its lower half and staminate spikelets on the remainder. This, as Fig. 13 shows, is a characteristic of ears borne in high positions on the stalk. If our reconstruction is valid should not prehistoric ears also bear staminate spikelets? A re-examination under the microscope shows that at least some of them once did and that these have since been lost in handling. Some of the ancient cobs, including the one illustrated in Plate II C, have stumps, previously unnoticed, of a slender stem upon which the staminate spikelets were undoubtedly borne. Thus our genetically reconstructed ancestral form has taught us to look for a characteristic in prehistoric ears which we had previously overlooked. It has also shown us the significance of ears bearing terminal male spikes which are still found in certain races of maize in the countries of Latin America: the races Nal-Tel and Chapalote of Mexico (Wellhausen *et al.*, 1952), Pollo of Colombia (Roberts *et al.*, 1957), and Confite of Peru (Grobman *et al.*, 1961).

In bearing both staminate and pistillate spikelets these ears of pod-popcorn also resemble the lateral inflorescences of *Tripsacum*, a perennial grass and a wild relative of maize (Plate II C). This resemblance has in turn called attention to additional characteristics in which the reconstructed maize resembles *Tripsacum*: (1) the flowering of the pistillate spikelets before the staminate in both lateral and terminal inflorescences; (2) the many-stalked condition; (3) the small, hard, pointed seeds.

Actually this reconstructed maize might easily be classified as an annual form of *Tripsacum* or conversely, since maize was the first of the two to be given a Latin name, *Tripsacum* could be classified as a perennial form of the genus *Zea* to which maize belongs and which until recently has been represented by the single species, *Zea mays*. These unexpected results of combining popcorn and pod corn—the production of a counterpart of maize's wild relative, *Tripsacum*—we regard as additional evidence that our reconstruction has validity.

One of the factors involved in evolution in nature is 'genetic

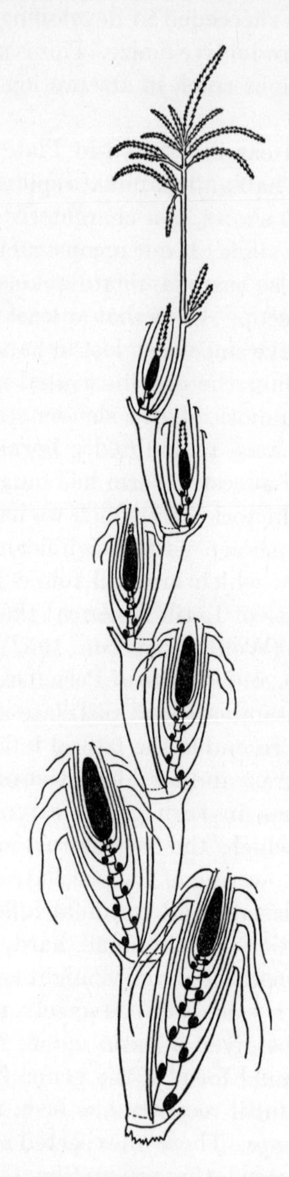

Fig. 13. Diagrammatic longitudinal section based on data from several plants of a many-eared stalk showing how the position of the ear on the stalk affects its characteristics. The higher the ear, the smaller its size, the fewer its husks, and the more likely it is to bear both male and female flowers. (Drawing by W. C. Galinat.)

drift' a term applied to changes in gene frequencies resulting from the random sampling of gene populations. Sewall Wright has given particular attention to the mathematics of this phenomenon. Population geneticists are not in complete agreement on the importance of this factor in evolution in nature but there can be little doubt that under some circumstances it is important. Populations which expand and contract greatly in numbers, such as those of certain species of insects, are especially subject to the influence of drift. Many genes are lost and others increased in frequency when the breeding population becomes greatly reduced in size as do some insect populations during over-wintering.

Genetic drift must almost certainly be an important factor in the evolution of cultivated plants in which gene populations are broken up rapidly and repeatedly. Vavilov, Russia's distinguished student of diversity in cultivated plants, recognized the phenomenon, although he did not call it genetic drift, and he employed it systematically as one means of determining centres of origin. He concluded from his world-wide studies that recessive genes, having low frequencies at the centre of a plant's origin, may attain high frequencies at the periphery of its spread simply as the result of random sampling or as Vavilov called it, the 'emancipation of recessives'.

In maize, experiments by Sprague (1939) have shown that ten to twenty plants are required for adequate representation of genetic diversity in an open-pollinated variety. Since the number of ears saved for seed by Indian maize growers, with only limited amounts of land at their disposal, is often smaller than this and, indeed, since new maize populations are sometimes established by growing the progeny of a single ear, it follows that there must often have been genetic drift—changes in gene frequencies resulting from small breeding populations.

An example of genetic drift in maize is the occurrence in Asia of varieties of maize with waxy endosperm (Collins, 1909, 1920; Kuleshov, 1928; Stonor and Anderson, 1949). Waxy endosperm is a simple Mendelian character in maize which affects the chemical composition of the starch, which in waxy varieties is composed exclusively of amylopectin and in non-waxy varieties of a mixture of amylose and amylopectin. Waxy varieties are unknown in pure form among the indigenous races of maize of America but the waxy character itself has been discovered in non-waxy varieties: in a New England flint maize (Mangelsdorf, 1924) and in a South American variety (Breggar, 1928). Bear (1944) reports that waxy endosperm is not an uncommon mutant in Corn Belt Dent varieties, he having found three separate mutations to waxy in three consecutive years in a total population of some 100 000 selfed ears.

The fact that waxy maize occurs so commonly in a part of the world which also possesses waxy (glutinous) varieties of rice, sorghum, and millet can be attributed to artificial selection. The people of Asia, being familiar with waxy varieties of other cereals and accustomed to using them for special purposes, recognized the waxy character in maize after the cereal was introduced into Asia following the discovery of America and purposely isolated varieties pure for the waxy condition. But the fact that waxy endosperm came to their attention in the first place is probably due to genetic drift. The gene for waxy endosperm which has a low frequency in American maize apparently attained a high frequency in certain samples of Asiatic maize. Indeed the practice, reported by Stonor and Anderson (1949) of growing maize as single plants among other cereals, would promote self-pollination and, in any stock in which the waxy gene occurred, would inevitably lead in a very short time to the establishment of pure waxy varieties whose special properties people, accustomed to the waxy character in other cereals, could hardly fail to recognize.

The selection and preservation of waxy maize in Asia exemplifies one of the most important features of genetic drift; the fact that it may interact with selection to confer greater adaptive plasticity than the species would otherwise possess. It may even produce genotypes which, being inherently inferior to the norm, must, to survive, find niches in which they may eventually become quite well adapted through the accumulation of modifying factors (Dobzhansky, 1955).

The development of waxy varieties of maize in Asia illustrates these steps to perfection. Waxy endosperm is inherently a defect in metabolism and its low frequency in most maize populations in the face of recurring mutations indicates that it is acted against by natural selection. In the small breeding populations employed in Asia, the waxy gene reached a high frequency in certain samples and because the people had special uses for cereals with waxy starch there was already at hand a niche for the new type in which it survived and flourished. More recently this niche has been greatly expanded by the development in the United States of new industrial uses of waxy maize and the establishment of a special market for it.

Other types of maize which differ from the 'wild' genotype primarily by single genes probably also had their origins through the interaction of genetic drift and artificial selection. These include both the flour and the sweet corns. Flour corn, the product of a locus on chromosome 2, found a special niche because it was much more easily chewed when parched than flint corn, the 'wild' type. Sweet corn, which involves a gene on chromosome 4, first found its special niche in the preparation of the native South American beer, *chicha*, to which, by virtue of its greater sugar content, it imparted a higher alcoholic potency, a characteristic prized no less in primitive societies than in modern civilizations. Like waxy corn, sweet corn eventually found a much larger niche in the United States, where used as green corn or 'roasting ears' and as canned and frozen corn, it has become a highly popular vegetable and is grown on an extensive scale.

Both floury and sugary maize are, like waxy, inherently defective. The sugary endosperm which sometimes occurs as a mutant in field-maize varieties is virtually a lethal condition. These types exist as cultivated varieties today because, having found niches in which they can survive, their deleterious effects have been counteracted by the accumulation of favourable modifying factors.

A somewhat different example of genetic drift is furnished by certain gene frequencies of the maize of the United States. Some years ago we test-crossed a number of varieties of maize of the countries of this hemisphere for the genes, *Pr* and *I*, which are concerned with the production of colour in the aleurone, and for a cross-sterility gene, *Ga*. The frequency of these genes in Mexican maize, from which the maize of the United States has been derived, was found to be 67, 50, and 56 per cent respectively. Among inbred strains isolated from commercial varieties of the

United States, the percentages of these same genes proved to be 98, 0, and 0. It is clear that the maize of the United States, the world's most extensive centre of maize production, is not at all typical of maize in general and represents an almost unique sample of the diversity which exists in this species. With respect to the loci under consideration it is an example of the decay of variability, one of the consequences of genetic drift.

There can be no doubt that natural selection acting in a man-made environment has been an important force in the evolution of maize under domestication and this has probably been especially true after maize became highly heterozygous as the result of hybridization between races and between maize and its relatives, teosinte and *Tripsacum*.

A number of characteristics of maize which are probably the product of natural selection might be mentioned: cold resistance of the high-altitude Peruvian maize, the resistance to the rust fungus of high-altitude varieties of Mexico and to *Helminthosporium* of tropical varieties of Colombia, the high chromosome-knob numbers of lowland tropical maize, the long mesocotyls of certain Hopi varieties which have been subjected for many generations to deep planting (Collins, 1914). All of these characters are probably the result of natural selection and it is doubtful that man participated to any substantial degree in their development except as he exposed maize to the environments in which these traits contributed to survival.

Natural selection, sometimes abetted by human selection, may be responsible for other characteristics: the early maturity of Canadian varieties, the long tight shucks providing protection against ear-worm and weevil damage of some varieties of the southern United States, the large pollen grains of varieties selected for their long ears. These may all be examples of the interaction between natural and human selection.

The effectiveness of natural selection in maize is well illustrated by the rapidity with which open-pollinated varieties become adapted to new environments. Kiesselbach and Keim (1921) showed some forty years ago that in Nebraska, which had then been settled only about a century, the maize varieties of the state had become so different in their adaptation that varieties collected from only a few hundred miles away were less productive than those from the county in which the yield tests were made. There are many other cases cited in the literature, but less precisely documented, of maize becoming rapidly modified in new environments.

Several students of maize, especially Kempton (1936) and Weatherwax (1942), have regarded cultivated maize as a product of the plant breeding skill of the American Indian. And it is certainly true that in the hands of Indian cultivators maize had reached a high state of development when America was discovered. All of the principal commercial types of maize recognized today, dent, flint, flour, pop, and sweet were already in existence when the first Europeans arrived and until hybrid maize was developed the modern maize breeder, for all of his rigorous selection, had made little progress in improving productiveness over that of the better Indian varieties.

It is doubtful, however, that the Indian was an accomplished plant breeder in the sense of visualizing a new type of maize and selecting toward it. Indeed, man in primitive cultures seems often to accept, with little or no effort on his part to change, the populations which other evolutionary forces have presented to him. If he recognizes degrees of excellence he is all too likely to consume first that which he considers best and to use for seed whatever remains. That cultivated plants have improved in spite of these disgenic practices testifies to the powerful force of natural selection interacting with other genetic agents of evolution.

The apparent absence of artificial selection in the early stages of domestication is well illustrated by several thousand years of evolution of maize as it is revealed by archaeological remains discovered in Bat Cave in New Mexico (Mangelsdorf and Smith, 1949). Although these show an increase in the average size of the cob and kernel during this period what actually occurred was a great increase in variability; kernel size, for example, becoming both larger and smaller than it was at the outset. The small primitive maize which was present at the beginning did not become extinct even at the end as it might have if rigorous selection in favour of large ears had been constantly practised.

At some stage in the development of Indian agriculture, however, artificial selection began and it is widely practised today.

The better Hopi maize growers never save seed from an ear which shows evidence of mixture (Wallace and Brown, 1956). Among the pure-blooded non-Spanish speaking Indians of Guatemala, rigid selection for type of seed is often practised (Anderson, 1947). Also in Guatemala there is a high correlation between the percentage of *indigenas*, pure Indians, and a number of distinct races of maize (Wellhausen *et al.*, 1957). In Peru selection has become an art and many different races are maintained in a state of uniformity, at least with respect to colour, by rigorous selection. The large-seeded flour corns of Peru as well as the sweet corns used in preparing *chicha* are the products of human selection interacting with mutation and genetic drift.

Scores of modern maize varieties owe their origin to rigorous selection for a specific type. Those which have furnished some of the inbred strains most widely used in the production of hybrid maize are of particular interest. In developing the famous Reid's Yellow Dent, Robert Reid of Illinois selected for an early maize with cylindrical ears bearing eighteen to twenty-four rows of kernels. George Krug, also of Illinois, selected for ears heavy for their size with lustrous kernels borne on good stalks. Isaac Hershey, the originator of Lancaster Surecrop, mixed together a number of varieties and selected for earliness and freedom from disease (Wallace and Brown, 1956).

The response, often spectacular, of modern maize to selection is nowhere better illustrated than in the famous Illinois experiments on selection for chemical composition. In fifty generations of selection, oil content, initially 4·7 per cent, has been raised to 15·36 per cent and reduced to 1·01 per cent in the high and low lines respectively. Protein content, initially 10·92 per cent, has been raised to 19·45 per cent in the high-protein line and reduced to 4·91 per cent in the low-protein line. That there is still genetic variability for these characteristics is shown by two generations of reverse selection which have been effective in three of the four lines (Woodworth *et al.*, 1952).

The fourth principal factor involved in the evolution of maize under domestication has been hybridization. This has included hybridization between races, both primitive and highly developed, and hybridization of maize with its two relatives, teosinte (Plate III) and *Tripsacum*.

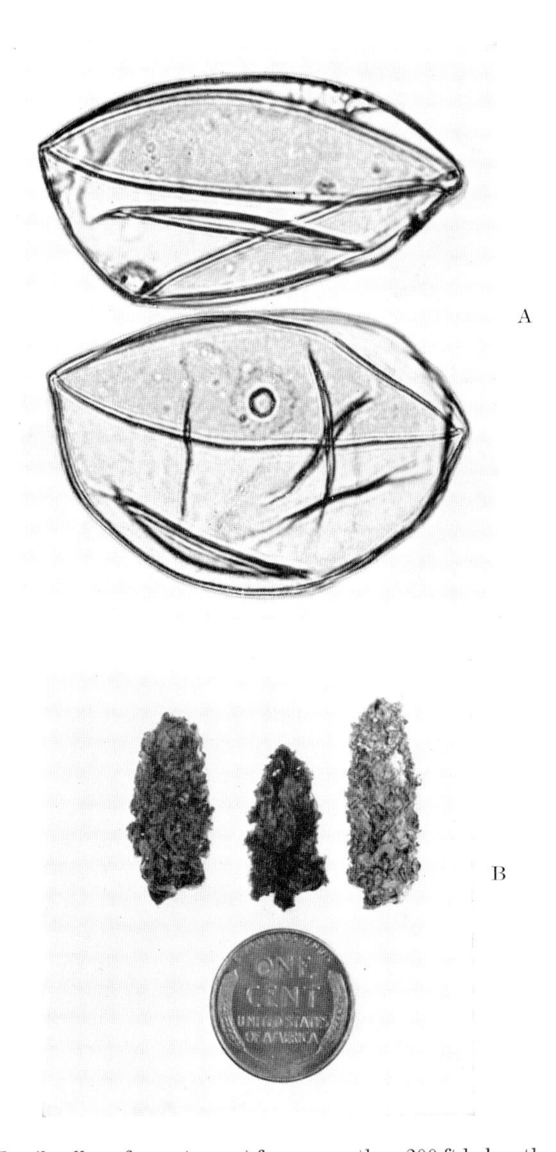

Plate I A. Fossil pollen of corn (upper) from more than 200 ft below the present site of Mexico City compared with a pollen grain of modern corn (lower) at the same magnification. In spite of some 80 000 years difference in their age these two pollen grains are virtually identical in their characteristics and they show that the ancestor of corn was corn and not one of its two American relatives, teosinte or *Tripsacum*. From Barghoorn *et al.* (1954).
B. Three cobs of prehistoric corn from Bat Cave compared with a one-cent piece. Radiocarbon determinations of associated charcoal date these at 5600 years. Natural size.

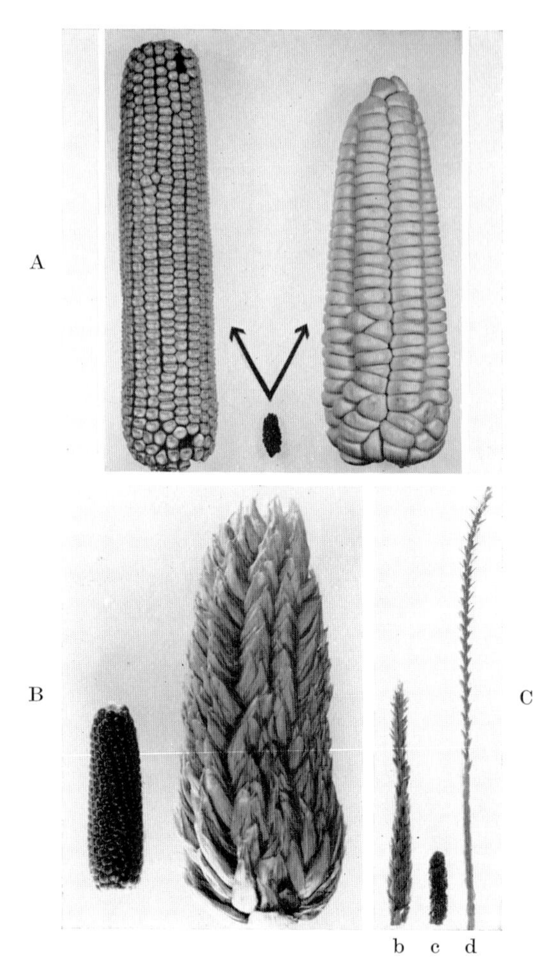

Plate II A. One of the Bat Cave cobs compared to a modern ear of Corn Belt dent (left) and a large seeded Peruvian flour corn (right). Extremely rapid evolution has been involved in producing such drastic changes even in 5600 years, the estimated difference in their age.

B. An ear of Argentine popcorn (left) and of present-day pod corn (right). Popcorn is primitive in having small hard seeds. Pod corn is primitive in having its seeds enclosed in pods or chaff as do the other cereals. These two types were crossed in order to combine their primitive characteristics.

C. The reconstructed ancestor of corn, an ear of pod-popcorn (b), compared with a modern ear of dent corn (Plate IIA) and a prehistoric cob of La Perra Cave corn (c). The dent corn weighs 317 g, the reconstructed ear 1·99 g. The reconstructed ear has female flowers below and male flowers above and in this respect resembles a spike of *Tripsacum* (d), a wild relative of corn. The La Perra cob (C) also once had a male portion which has been lost; only the stump of its stem still remains. Without its kernels and male spike, the reconstructed ear would weigh 0·87 g; only slightly more than the La Perra cob which weighs 0·52 g.

A

B

Plate III A. Plant of teosinte, *Zea mexicana*, the closest relative of maize which occurs in Mexico, Guatemala, and Honduras often as a weed in the corn fields. The hybridization of maize with teosinte has resulted in new genetic recombinations and has also had mutagenic effects.

B. Pistillate spikes of teosinte. These are distichous and bear single spikelets. They are fragile and disarticulate when mature. The spikelets are sessile and the kernels are enclosed in shells consisting of rachis segments and lower glumes. It is this last named characteristic which when introduced into maize through introgression is responsible for the lignification of the tissues of the rachis and lower glumes. Introgression of *Tripsacum*, another relative of maize, has similar effects.

Plate IV. Prehistoric cobs from Swallow Cave illustrating an evolutionary series. a and b, pure maize from the lower levels. c, a tripsacoid cob exhibiting teosinte introgression and hybrid vigor. d, an eight-rowed race introduced from South America exhibiting *Tripsacum* introgression and hybrid vigor. e and f, cobs from upper levels which may represent the product of four kinds of heterotic effects resulting from: (1) interaction between the genes from different races of maize; (2) heterotic interaction between *Tripsacum* and maize genes; (3) heterotic interaction between teosinte and maize genes; (4) interaction between teosinte and *Tripsacum* genes.

It has been known for some years that the Corn Belt Dent of the United States, the world's most widely grown maize, is a hybrid of two distinct races which were being grown by the Indians when European colonists arrived in America. One of these, commonly called Gourd Seed, was a late-maturing dent type grown in the South and the other, an early-maturing flint corn, was the principal variety of the north. The two races became crossed when farmers, in what is now the Corn Belt, adopted the practice of replanting the missing hills of the southern maize with seed of the early-maturing northern flint corn. One result of this practice was that the late-planted early-maturing flint maize flowered at the same time as the earlier-planted late-maturing dent maize. Maize being largely cross pollinated, the two races hybridized freely and from the hybrid population a relatively stable new race was selected (Anderson and Brown, 1952; Wallace and Brown, 1956).

The rôle of racial hybridization in the evolution of maize is especially well illustrated by the work of Wellhausen *et al.* Of the twenty-five races recognized in Mexico, seventeen were regarded as the products of hybridization either between pre-existing races or between maize and teosinte.

Since the Southern Dent maize of the United States has obvious affinities with a Mexican race, Tuxpeño, Wellhausen *et al.* were able to trace back one half of the ancestry of Corn Belt Dent to Mexican races originally from South America. More recently Grobman *et al.* (1961), in classifying and describing the races of maize of Peru, greatly extended the genealogy of Corn Belt Dent. They not only traced back several of the Mexican races to a popcorn of Peruvian origin but also showed that Northern Flint, the other parent of Corn Belt Dent, has a complex ancestry tracing back through Mexican races to another race of popcorn of South American origin.

No less important than racial hybridization in the evolution of maize is its hybridization with its relatives, teosinte and *Tripsacum*. It is doubtful whether among varieties of maize now in existence there is any that is completely free of introgression from teosinte or *Tripsacum*. Perhaps the only 'pure' maize known today is the prehistoric maize found in archaeological sites, for example, the early Bat Cave maize described in a preceding part of this chapter.

Although several students of maize (Brieger *et al.*, 1958; Randolph, 1955) have expressed scepticism about the hybridization of maize with teosinte and *Tripsacum*, the evidence for such hybridization is extensive and comes not from one source but from several: archaeology, morphology, genetics, cytology, and systematics. The various kinds of evidence have been described in detail in a recent paper (Mangelsdorf, 1961). Here it will suffice to state that genetic experiments with maize-teosinte and maize-*Tripsacum* hybrids have shown that the introgression of teosinte or *Tripsacum* into maize produces certain easily recognizable morphological effects, especially those concerned with the lignification and induration of the tissues of the rachis and lower glumes. Consequently maize with highly indurated tissues of the rachis and the lower glumes is suspected at once of being the product of introgression from either teosinte or *Tripsacum*. When such maize exhibits additional characteristics of teosinte or *Tripsacum* such as distichous spikes and single spikelets the evidence of introgression is regarded as virtually conclusive.

Tripsacoid specimens—cobs resembling segregates of maize-teosinte or maize-*Tripsacum* hybrids—have been found in numerous archaeological sites in Mexico and the south-western United States (Galinat *et al.*, 1956; Galinat and Ruppe, 1961; Mangelsdorf and Lister, 1956; Mangelsdorf *et al.*, 1956; Mangelsdorf and Smith, 1949). They occur also among living varieties in virtually all of the countries of Latin America. In these the external morphological characteristics of induration of the tissues are often associated with internal cytological features, especially knobs on the chromosomes (Brown, 1949; Mangelsdorf and Cameron, 1942; Wellhausen *et al.*, 1952). Tripsacoid characters found in the varieties of Mexico and Central America can be attributed to contamination with teosinte which grows in and around the maize fields often in great profusion (Collins, 1921; Kempton and Popenoe, 1937; Mangelsdorf, 1952) but tripsacoid cobs occurring in varieties of South American countries where teosinte is unknown may have a different origin. These may be the product of hybridization of maize with South

American species of *Tripsacum* which occur in all the countries of northern South America except Peru (Cutler and Anderson, 1941). The fact that maize can hybridize more freely with *Tripsacum* than had previously been supposed (Farquharson, 1957; Galinat, unpublished) and that in the hybrids there is crossing over between the chromosomes of the two genera, (Maguire, 1957) tends to support the conclusion (Grobman *et al.*, 1961; Mangelsdorf, 1961) that there has been introgression of *Tripsacum* into maize in South America.

The introgression of teosinte and *Tripsacum* has played several rôles in the evolution of maize. One obvious result of such introgression has been the production of new genotypes through genetic recombination, some of which have been superior to the original maize in various respects. It seems improbable, for example, that the large modern ear of maize could have evolved until after hybridization with teosinte which contributed genes for induration and lignification thereby providing strength for a greatly enlarged structure. On this point I once expressed the following opinion (Mangelsdorf, 1950):

> The elements of strength necessary to support this greatly enlarged inflorescence have come from teosinte, which contributed genes for hardness and toughness when it is hybridized with corn. Teosinte is to the modern ear of corn what steel is to the modern skyscraper.

Other characteristics which may have been transmitted to maize through gene recombination are resistance to heat and drought, to certain diseases, and even to insect damage. Horowitz and Marchioni (1940) have suggested that the resistance to grasshoppers of *Maiz amargo* is due to introgression from *Tripsacum*. Cervantes *et al.* (1958) have recently found a remarkable correlation between the estimates of teosinte introgression made by Wellhausen *et al.* (1952) and susceptibility to a virus disease, 'stunt', which is transmitted by a species of leaf hopper.

The second effect of introgression from teosinte and *Tripsacum* has been to increase the mutability of maize. For some years we have been introducing chromosomes of several varieties of teosinte into an inbred strain of maize through repeated backcrossing. The original purpose of this experiment was to determine, by comparing the modified strains with the original, what kind of genes are carried by the introduced chromosomes and whether different varieties of teosinte are similar in the genic constitution of their chromosomes. An unexpected result of the experiment, now overshadowing its original purpose, has been the discovery that the introduced chromosomes have mutagenic effects (Mangelsdorf, 1958). In some lines mutations have occurred in 7 to 9 per cent of the plants. The mutations so far detected have all been deleterious but there is reason to believe that beneficial mutations also occur especially if some of the mutations are the result, as they may well be, of deficiencies resulting from unequal crossing over, a phenomenon which would be expected to give rise also to minute duplications. The duplications would not usually be detectable but might represent useful new building blocks of evolution.

It is of particular interest to note that chromosomes extracted from tripsacoid South American varieties have mutagenic effects similar to those introduced from teosinte and that some of the mutations which occur are genetically identical with those produced by teosinte chromosomes (Mangelsdorf, 1958).

A third effect of teosinte and *Tripsacum* introgression is to introduce additional heterosis into maize thus contributing to making it one of the most heterotic of cultivated plants propagated by seed, in which the heterosis is not due to allopolyploidy. Grobman *et al.* (1961) have pointed out that some varieties of modern maize may exhibit five different kinds of heterosis: (1) resulting from the interaction of the genes of different races of maize; (2) involving interaction between teosinte and maize genes; (3) produced by the interaction between *Tripsacum* and maize; (4) resulting from the interaction of genes from teosinte and *Tripsacum*; (5) involving different species or geographic races of *Tripsacum* which have contributed genes to modern maize.

The archaeological specimens from Swallow Cave in northern Mexico illustrated in Plate IV show the effects of the three kinds of hybridization which have played a part in the evolution of maize under domestication. The small cob at the left, the oldest, is similar to the earliest cobs from Bat Cave in New Mexico. It is a prototype of one of the Ancient Indigenous races of Mexico, Chapalote. The next cob shows the extent to which evolution under domestication had increased the size of the cob before hybridization of any kind had occurred. The third cob, with its prominent indurated glumes, shows the effects of hybridization with teosinte. The fourth represents the introduction of the

Fig. 14. Environmentally induced and genetically controlled variation in the corn plant. *Left:* a plant of pod-popcorn as it might have grown in nature in a poor site in competition with other natural vegetation. *Second:* the same grown under primitive agricultural conditions. *Third:* the same in a fertile site free of competition with weeds. *Fourth:* a popcorn plant which has lost the pod corn gene. *Fifth:* human selection for larger ears has tended to eliminate the secondary stalks reducing them to 'suckers' as in this typical plant of a New England Flint corn. *Right:* this trend has been carried still further in Corn Belt Dent corn which is usually single stalked, commonly bearing a single large ear in the middle region of the stalk. The middle position of the ear has both mechanical and physiological advantages over a terminal position and probably accounts for corn's superiority over other cereals in its capacity to produce grain. (Drawing by W. C. Galinat.)

South American race, Harinoso de Ocho, which we now have reason to believe is a product of hybridization with *Tripsacum* in South America (Grobman *et al.*, 1961). The two remaining cobs illustrate the explosive evolution which followed the hybridization of these two races, one carrying blocks of genes from teosinte and the other from *Tripsacum*.

Figure 14 illustrates some of the principal environmentally induced and genetically controlled changes which are believed to have occurred during domestication. The first three plants illustrate the genetically reconstructed ancestral form, a pod-popcorn, as it would be expected to develop in three different environments. The first plant, a short, single-stalked plant with a slender, unbranched tassel bearing both male and female flowers and no ears, is intended to represent the wild maize plant growing in nature in a site of low fertility and in rigorous competition with other natural vegetation. Such a plant would barely reproduce itself.

The second plant represents this same genotype grown in a more favourable natural site or under primitive agricultural conditions. Here it is still single-stalked but under these somewhat better conditions is capable of producing a branched tassel and a single small ear borne high upon the stalk. The third plant represents the genetically reconstructed ancestral form grown under modern agricultural conditions with an abundance of fertilizer and free from competition with weeds. Under these conditions it has several stalks each bearing several small ears. Plants like these might also have occurred sporadically in the wild under unusually favourable natural conditions.

The ability of the wild maize plant to respond impressively to freedom from competition with weeds and to high levels of fertility is undoubtedly one factor which led to its domestication. This ability to take full advantage of the improved environment usually afforded by an agricultural system is one of the characteristics found in almost all highly successful domesticated species. There are many wild species which do not have this trait; they cannot stand prosperity.

Since the maize plant is genetically plastic as well as responsive to an improved environment, domestication may soon have brought other changes, which are illustrated in the last four plants in Fig. 14. One of the most important of these was a mutation at the pod-corn locus on the fourth chromosome. This single genetic change had numerous effects. It reduced the glumes, which in wild maize completely surrounded the kernels, and the energy released from chaff production now went into the development of a larger cob, which in turn bore more and larger kernels. The mutation also lowered the position of the lateral inflorescences and this had collateral effects of several kinds which can be understood by referring again to Fig. 13. This shows that: (1) The lower the ear, the stronger the stalk at the position at which the ear is borne and the greater its capacity for supporting large ears. (2) The lower the ear, the more likely it is to bear only pistillate spikelets which develop kernels when pollinated. (3) The lower the ear, the longer the shank, the branch on which it is borne, and this in turn has a number of important secondary effects: the longer the shank, the more numerous its nodes or joints and the husks which arise from them; the greater the number of husks, the more completely the ear is enclosed and the less capable it is of dispersing its seeds.

In brief, a rather simple change but a very important one, the lowering of the position of the ear (comparable, perhaps, to moving the engine of a primitive airplane from a position behind the wings to one in front of them), has separated the sexes, and made for a larger strictly grain-bearing ear which is completely protected by the husks and is no longer capable of dispersing its seeds. Thus, a mutation at a single locus on chromosome 4 has made the maize plant less able to survive in nature but much more useful to man.

The last two plants in Fig. 14 show some of the changes which human selection acting upon the remarkable diversity created by mutation, genetic drift and hybridization has subsequently effected. Selection for large ears has tended to eliminate the secondary stalks and to reduce the number of ears per stalk. The fifth plant in Fig. 14 represents a typical New England Flint maize in which the secondary stalks have been reduced to low tillers, known to the farmer as 'suckers', which in days of cheaper labour were often removed under the erroneous impression that their removal was a kind of beneficial pruning operation. The last plant represents a typical Corn Belt Dent maize which is predominantly single-stalked and often bears only one ear in approximately the middle region of the stalk.

At the time that the drawing in Fig. 14 was made we had not yet actually produced the plants depicted on the left but knowing from long experience the ways in which the maize plant responds to unfavourable growing conditions we were reasonably certain that we could produce these types by restricting soil fertility. We first attempted to do this by growing plants thickly in the row, three inches apart instead of the usual twelve, but because one of the adjoining rows turned out to have a poor stand the competition between plants was less intense than had been intended and we failed to produce the first plant in the series; we did, however, obtain one plant which was a close counterpart of the second.

The following year we grew the reconstructed ancestral form in a simulated wild habitat, a weedy fence row in which the maize plants were forced to compete, without the benefit of fertilization or cultivation, with other vegetation including the aggressive perennial couch grass, *Agropyron repens.* Under these conditions the reconstructed ancestral form was even more stunted than the diagram depicts it but its botanical characteristics were exactly those which we had expected. It had neither tillers nor ears and its unbranched terminal inflorescence bore staminate spikelets above and pistillate below. Subsequent experiments, not yet published, suggest that plants of this type compete successfully with other vegetation because the major part of their leaf area is devoted not to nourishing the grain bearing inflorescences as in modern maize but to nourishing the root system.

Four evolutionary forces: mutation, genetic drift, selection, and hybridization interacting to a degree seldom encountered in nature and accelerated and intensified by the activities of man have produced in the maize plant evolutionary changes so profound and in so short a time that a paleontologist seeing the species only at the beginning and the end might well suspect that evolution under domestication is cataclysmic and the product of violent saltation. When we observe the intervening steps, as in some instances, thanks to well-preserved archaeological remains, we can, we see that this is not true and we see also that evolution under domestication is a close counterpart of evolution in nature, the product of what Wright has called the 'interplay of directed and random processes'. Man has played an important part in it but perhaps more important still is its impact upon him. It can be said with some degree of truth that man's rise from a state of savagery to one of civilization began when he accidentally set in motion and himself became involved in the genetic forces of evolution acting upon animals and plants under domestication. And in America it was the maize plant—at the outset by no means the most promising of the species chosen for cultivation—which responded in the most spectacular fashion to the evolutionary forces operating under domestication and which as a consequence became not only the basic food plant of all advanced cultures and civilizations of the New World but also in the words of a perceptive student of the history of maize 'the bridge over which English civilization crept, tremblingly and uncertainly, at first, then boldly and surely to a foothold and a permanent occupation of America'.

III

THE DEVELOPMENT OF THE CULTIVATED SORGHUMS

by H. DOGGETT

THE cultivated *Sorghums* are grasses of the tribe Andropogoneae, ranging in height from less than three to over fifteen feet. They are known by a variety of names, including dura, jowar, guinea-corn, kafir corn, milo and the great millet. Their grains are used for food and brewing, the vegetative parts make a good forage and silage, and the stalks are used for fencing, building, and basket-making. They come fourth in the order of world acreage of cereals, being exceeded by wheat, rice and maize.

Sorghum was originally classified under *Holcus* Linn. (Snowden, 1936) and was established as a separate genus by Moench. Hackel and others grouped *Sorghum* under *Andropogon sorghum*, regarding it as a subgenus. Stapf (1917) in the *Flora of Tropical Africa* raised *Sorghum* to full generic rank, and grouped it with *Cleistachne*, *Vetiveria*, and *Chrysopogon* as the Sorghastrae. Within the genus *Sorghum*, he made two sections, Eusorghum and Sorghastrum. These were distinguished by the type of panicle branching, and the status of the pedicelled spikelets. Snowden (1936) kept Sorghastrum as a distinct genus, as Nash had done originally prior to Stapf's work, and separated within *Sorghum* the two main sections Parasorghum and Eusorghum. Parasorghum has bearded nodes, and simple panicle branches, and Snowden's Eusorghum was that of Stapf with the Parasorghums removed from it. Eusorghum is distinguished by the glabrous or finely pubescent nodes, the sub-division of the primary branches of the panicle, and the lateral and terminal racemes.

In the last fifteen years, Garber (1950) has made a study of the genus, including cytological and taxonomic investigations, as well as some hybridization work. Celarier has contributed confirmatory supplementary evidence, especially on the cytology. Garber used the group name Sorghastrae, and retained in it the two main genera of *Sorghum* and *Cleistachne*. Within the genus *Sorghum*, he split off six sub-genera, as follows:

Group Sorghastrae

Genus *Sorghum*:

(a) subgenus Eusorghum, which is the same as Snowden's section Eusorghum.
(b) Chaetosorghum.
(c) Heterosorghum.
(d) Sorghastrum.
(e) Parasorghum.
(f) Stiposorghum.

Genus *Cleistachne*.

There are sharp distinctions between many of Garber's sub-genera. Garber's classification seems generally acceptable, though Celarier (1958*b*) has criticized it on some points. In this paper Garber's system will be used, and the world distribution of the *Sorghums* on his classification is set out in Figs. 15 and 16. In Fig. 15 Garber's Eusorghum group is sub-divided into diploid species and tetraploid species, an important consideration when following the spread of the cultivated grain *Sorghums* from Africa into India and the Far East.

The basic chromosome number within the genus *Sorghum* is 5. The chromosomes of the sub-genera Parasorghum and Stiposorghum are relatively large, about twice the length of those of the other sub-genera at Metaphase I (6 μ compared with about 3·5 μ), and the haploid number of 5 is found in these two sub-genera. They have tetraploid and hexaploid members. The haploid number of Eusorghum is 10, with both diploid and tetraploid members represented. Sorghastrum is similar, but with hexaploid members as well, and both Chaetosorghum and Heterosorghum have haploid numbers of 20.

The distribution of the genus *Sorghum* as delimited by Garber is world wide. At the one end, Sorghastrum is represented in America and Africa. At the other, both Chaetosorghum and Stiposorghum are confined to Australia. Parasorghum spreads

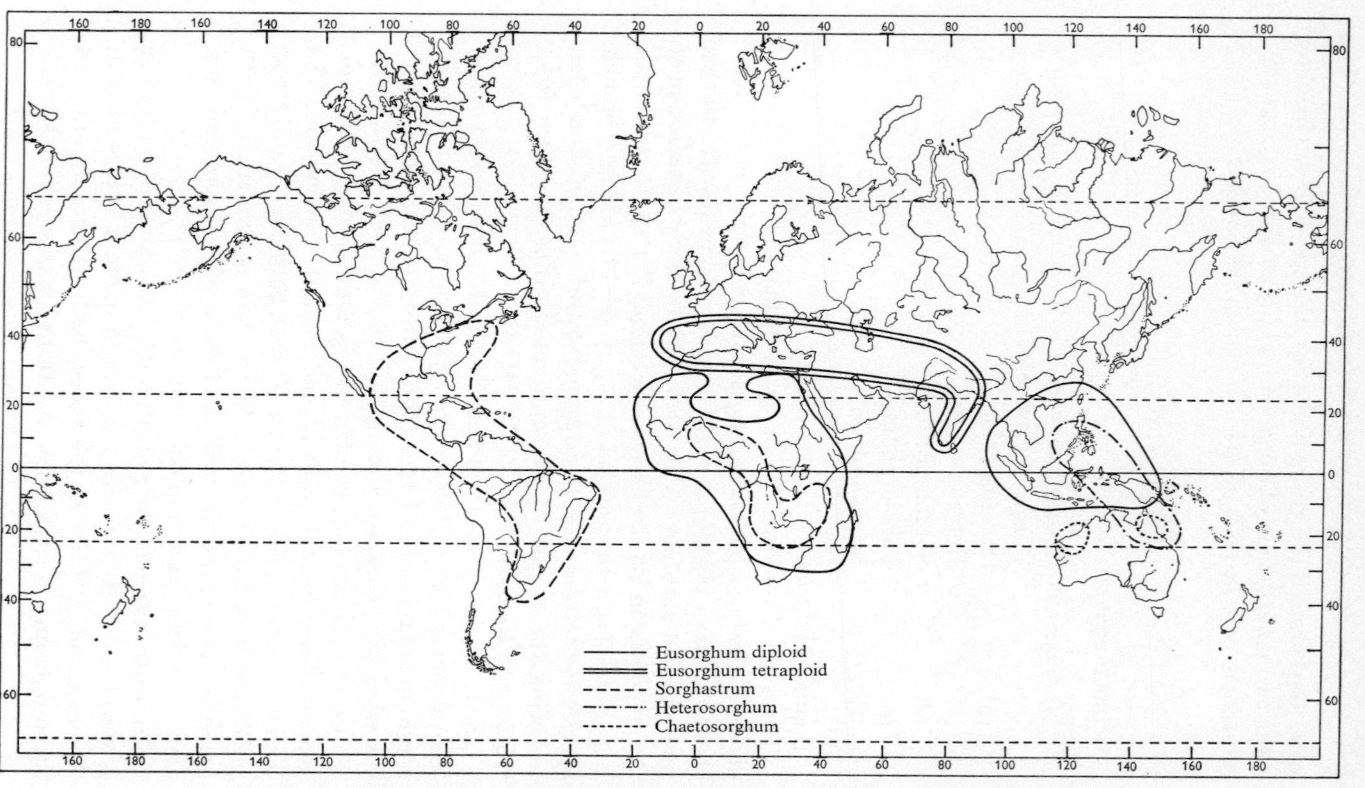

Fig. 15. Distribution of Eusorghum, Heterosorghum, Chaetosorghum and Sorghastrum.

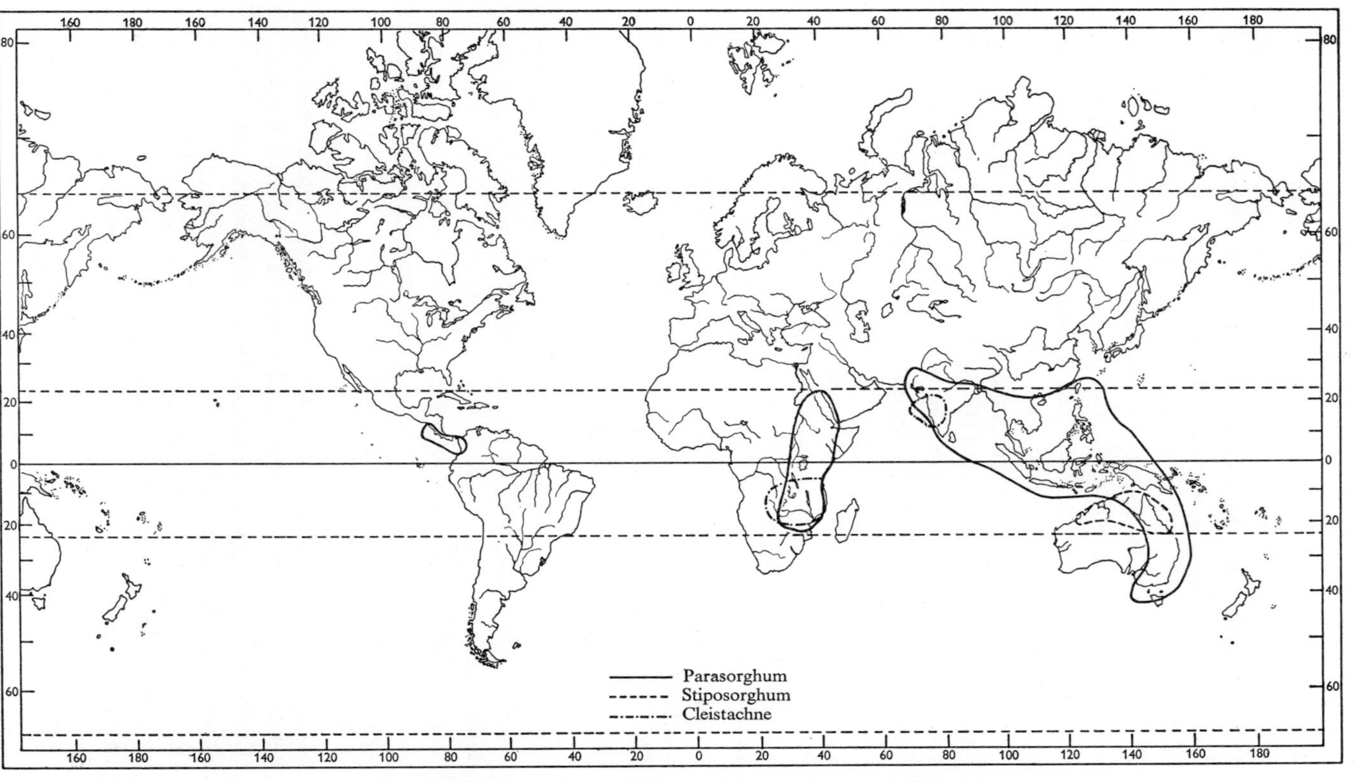

Fig. 16. Distribution of Parasorghum, Stiposorghum and Cleistachne.

from southern and eastern Africa through India and South-east Asia to Australia, and to western Guatemala and Mexico. A Parasorghum type, *S. trichocladum*, from the western side of Mexico and Guatemala shows similarities to the *S. australiense* of Australia. Heterosorghum is restricted to New Guinea, the Philippines, and Australia. Eusorghum is distributed through southern Africa to India, South-east Asia and the Philippines, but did not reach either Australia or America until carried there by man in the past few centuries.

The distribution of Eusorghum overlaps with that of Sorghastrum, Parasorghum, and Heterosorghum. No hybridization attempts between Eusorghums and Sorghastrum have been reported, but all the morphological evidence indicates that they are genetically widely separated, and we may assume that fertile hybrids between them do not occur. The evidence on the intersterility of the Eusorghums and Parasorghums is better. Several workers have attempted to cross representatives of the two groups, without success. I have tried the cross on several occasions using diploid and autotetraploid plants of both, in all combinations, and there are no signs of fertility between them. The only overlap between Eusorghum and Heterosorghum is in the Philippines, and although hybridization between the two has not been attempted, to my knowledge, there is no reason to expect that they will prove interfertile. On all the available evidence, therefore, we may regard the Eusorghums as a sharply defined group.

In considering the development of the cultivated Sorghums, attention may be confined to the Eusorghum section, and from now on I shall use the term 'sorghum' for the cultivated forms of the sub-genus Eusorghum. The diploid Eusorghums with a chromosome number of 20 show completely regular meiotic behaviour. There is a suggestion from a study of haploids (Endrizzi and Morgan, 1955) that they are ancient tetraploids, which would be expected since the basic chromosome number of the genus is 5. The tetraploid Eusorghums, $2n = 40$, are not quite regular in meiotic behaviour. (Endrizzi, 1957; Celarier, 1958*a*.) There are usually two to three quadrivalents per cell, and there is a low frequency of univalents.

The section Eusorghum is subdivided into two subsections,

Halepensia and Arundinacea. The Halepensia are characterized by the possession of elongated rhizomes, and are perennial wild grasses. Many members of this group are tetraploid, although some diploids occur. Members of the subsection Arundinacea have no rhizomes, they may be annuals, but are often perennial under favourable conditions, persisting by the proliferation of fresh buds from the bases of the old culms. They are all diploids, and it is to the subsection Arundinacea that the cultivated Eusorghums belong.

The wild Arundinacea (Snowden, 1955) have a natural distribution in Africa, over virtually the whole continent, with two collections of rather uncertain status in India. They are generally grasses of river or swamp margins, or weeds of cultivated and abandoned cultivated land, but some are found in arid conditions, usually in the beds or along the edges of seasonal watercourses. One diploid member of the Halepensia, *S. propinquum*, is found in South-east Asia, Indonesia, and the Philippine Islands. It has been collected in southern India and Ceylon, but was probably introduced there, and it seems to be a forest grass well adapted to moist habitats such as river banks. Very occasional collections of other diploid Halepensia have been made in India. The tetraploid Halepensia occupy a continuous area from southern and eastern India, Pakistan, Afghanistan, across Asia Minor and the Levant, and all round the Mediterranean, covering a range of habitats from jungle margins and clearings to dry, open country. In the Mediterranean region and between that sea and India, they seem to be predominantly weeds of cultivated land, abandoned cultivations and roadsides.

All the Eusorghums are interfertile, with one exception to be noted later. *S. propinquum* crosses readily with the African Eusorghums of the Arundinacea sub-section, and the progeny are fertile with regular meiotic divisions. I have only had later generation progeny of two crosses, a kafir × *S. propinquum*, and a Sudan grass × *S. propinquum*, and these have shown no obvious abnormalities. Much more work has to be done before the extent of barriers between the two can be assessed. Taxonomically they are distinct. *S. propinquum* has well developed, strong rhizomes, extremely small seeds, and a characteristic leaf shape, tapering towards the stem. It must have been isolated from the

African Eusorghums for a long period, and its behaviour in crosses with them gives an indication of the problems the systematic botanist faces in defining *Sorghum* species. The studies made of the Old World Halepensia show that they probably had an allopolyploid origin, but with the genomes involved being rather closely related, one genome being that of the Arundinacea sub-section. The distribution of the tetraploid Halepensia taken in conjunction with the cytological evidence makes it very probable that they arose from hybridization of an Arundinacea type with a diploid Halepensia type, followed by chromosome doubling. There are, however, other possibilities.

The tetraploid Halepensia (Hadley, 1958) cross with the diploid Arundinacea with low frequency, and the progeny of such crosses are often tetraploid owing to the functioning of an unreduced (or doubled) gamete from the Arundinacea parent. These progeny are fertile, and produce functional pollen. *S. almum* in fact arose in South America in this way in recent times.

Having restricted our search for the origins of cultivated *Sorghums* to the section Eusorghum, sub-section Arundinacea, and having noted the distribution of the wild *Sorghums* and their potential crossing relationships, we now turn to the distribution of the wild Arundinacea in Africa. One of the striking features of these wild grasses is their variability, and systematic botanists have split them into species, the most recent revision being that of Snowden (1955). Opinions differ on the status of these species, but without entering into this controversy we will treat Snowden's species as giving an estimate of the available variability. He lists sixteen species of wild Arundinacea Eusorghums in Africa, extending Stapf's (1917) original classification. Of these sixteen, fourteen have been collected from the north-east quadrant of Africa, using this term for that region north of latitude 10°S, and east of longitude 25°E. If varieties are taken into consideration, nineteen of the twenty-one varieties have been collected in this quadrant, with only two West African forms, *S. arundinaceum* and *S. vogelianum* missing from this region. For the western side of Africa, five of the sixteen species, or five of the twenty-one varieties have been collected, while for southern Africa, below latitude 10°S, six of the species or eight of the varieties have been collected. These figures are up-to-date on the present Kew material. Thus, it looks as though the greatest variability of the wild diploid Arundinacea occurs in the north-east quadrant of Africa.

Turning now to the cultivated forms, these have also been given specific names by Snowden, extending Stapf's classification. We may again regard Snowden's species and varieties as giving an estimate of variability. Our data are based on Snowden's 1936 book. This is perhaps unfortunate, but to bring it up to date it would be necessary to go through all the cultivated material at Kew, which I have not yet been able to do. The data are biased in favour of territories that were British possessions before 1939, as a request was sent to them to collect material.

Of the thirty-one Snowden species, twenty-eight occur in Africa; this places the greatest variability of the cultivated Eusorghums in Africa. We have already seen that the wild Arundinacea are confined, or almost confined, to Africa. We can regard Africa as the home of the cultivated Eusorghums with some certainty. Within Africa, twenty Snowden species occur in the north-east quadrant. This includes collections from Egypt, Eritrea, Somaliland, Abyssinia, the Sudan, Kenya, Uganda, Tanganyika and Zanzibar. The Sudan has eleven species, Tanganyika nine, and Eritrea five. Unfortunately, there are only eight collections from Abyssinia, but these yield three species and seven varieties. Somaliland, with only nine collections, has four species and six varieties. There is thus a good deal of variability in the Sudan, and in Tanganyika, and we may suspect that there is a great deal in the Abyssinia-Somaliland region.

On the western side of Africa, north of the Gulf of Guinea, including Tunis, Senegal, Gambia, French Guinea, Sierra Leone, Liberia, Ghana, Togoland, Nigeria, and the Cameroons, there are eleven Snowden species. Nigeria has eight species, Ghana four, Togoland and the Gambia three each, and Sierra Leone two. Very few specimens have been sent to Kew from the other West African areas. From Africa south of the Congo and Tanganyika, twelve species have been collected.

Of the twenty species in the north-east quadrant of Africa, eleven are not found on the western side of Africa. Of the eleven species on the western side, four are not found on the eastern side of Africa. Of the twelve species in southern Africa, four are not found in the north-east quadrant. It therefore appears that there is more variability in the north-east quadrant of Africa than elsewhere on that continent. An analysis of Snowden's varieties gives a similar picture. Eighty-seven have been collected in the north-east quadrant, and twenty-seven from West Africa north of the Gulf of Guinea. Of these twenty-seven, eight also occur in the north-east quadrant. Forty varieties have been collected in Africa south of the Congo and Tanganyika, of which twenty-one have also been collected in the north-east quadrant. One variety has been collected in all three areas.

Vavilov considered Abyssinia a centre of variability for *Sorghum*. The evidence on the material available to Snowden shows that the main variation is in the region of Africa north of latitude 10°S, and east of longitude 25°E. When we consider the coincidence of the greatest variability of the wild and cultivated *Sorghums* in this north-east quadrant of Africa, it seems that we must look there for the origin of the crop.

Helbaek (1959) has shown that primitive wheats were in cultivation at Jarmo in the uplands of Iraq-Kurdistan at the beginning of the seventh millennium B.C., and he states that the emmer (*dicoccum*) wheats reached Egypt in the 5th millennium B.C. These emmer wheats are also common in Abyssinia, though there is no evidence on when they reached there. Helbaek (1955), discussing a wheat spike from el Omari in Egypt, dated as late, or middle and late, Neolithic, says: 'The apex of the glume is perfectly preserved in seven out of eight spikelets, corresponding closely to this detail in emmer. Five apices recall the present Abyssinian type, by their narrowly converging vein tips.' It seems probable that the Abyssinian emmer wheats reached that country at an early date.

Murdock (1959), in his book on Africa and its peoples, cites a people in the Ethiopian region who were of Caucasoid origin, speaking the languages of the Cushitic sub-family of the Hamitic Stock. These Cushites were probably there by the third millennium B.C., and Murdock associates them, especially the Agau people, with the development of several crops during the third millennium. He considers that they did not originate the cultivated sorghums, but received and diversified them. Murdock would place the origin of cultivated sorghum in West Africa, near the headwaters of the Niger river, where he thinks that there was an independent origin of agriculture. To me, it seems more probable that the Cushites brought agriculture to Africa, possibly with emmer type wheat. These Caucasoid people presumably passed through or near Iraq-Kurdistan on their journey to Abyssinia. It is clear, however, that agriculture reached West Africa at an early date, and quite a lot of crop development took place over there. As far as sorghum is concerned, the balance of the evidence, such as it is, favours development over the east side of Africa, followed by spread to the west and further diversification there. Mauny (1953) states that the crop was grown in West Africa in Neolithic times.

We may suggest that the Cushites grew emmer wheat, and being accustomed to cereals, developed some of the local plants, including *Eleusine* millet, teff, flax and sorghum, for use in the areas less well suited to wheat. It is not of course necessary to assume that the Cushites had wheat at first, but it does seem likely. To a people practising primitive agriculture the wild Eusorghums would be obvious grasses to adopt and improve. Indeed, they might well have been weeds in their fields. I was pleased to find at Kew recently a specimen of *Sorghum sudanense* collected from the edge of a wheat field at Kiling on Jebel Marra in the Sudan, at an altitude of 7000 ft.

Murdock believes that a branch of the Cushites pushed down into East Africa, and occupied areas suited to their agriculture. These were local and scattered, the bulk of the region being peopled only by Bushmanoid hunters and food gatherers. Certainly there was a Neolithic agriculture in East Africa. Remains of terraces and irrigation furrows are to be found at various sites in Kenya and Tanganyika, and today small agricultural groups of the Topeth persist in Karamoja on some of the mountains, though their history is obscure. The Gumban cultures of Kenya had stone bowls, mortars and grinding stones which perhaps date from the third millennium B.C., according to Leakey (1931), and the Njoro River cave in Kenya has yielded pestles, bowls and grinding stones dated about 850 B.C. (Leakey and Leakey, 1950). The people who used these implements are likely to have had sorghum. I am certain that a connexion between the Abyssinian sorghum culture and that of parts of Tanganyika will one day be proved. Brooke (1958) gives a description of the growing and harvesting of sorghum in the Central Highlands of Ethiopia, at altitudes ranging around 5500 ft. There, the crop is the mainstay of the agriculture, and is found in conjunction with barley, sesame, field peas, fenugreek, teff, flax, emmer wheat and lentils. The account of the harvesting, seed selection, preparation of the threshing floor, communal threshing by the menfolk using long sticks, sweeping, winnowing with baskets, and storing which Brooke gives, describes exactly the present day practices of the Wasukuma in the Lake Province of Tanganyika. Their practices are quite different from those of the Nilo-Hamitic and Nilotic people who today occupy the region between the Wasukuma and Abyssinia.

Reverting to the development of cultivated sorghum, an important feature of Africa is of course the belt of tropical forest running from West Africa through the Congo to the Great Lakes, which is a formidable barrier to north–south movement. North–south movement was, in fact, restricted to the belt of savannah country between the East African Lakes and the Indian Ocean. Africa south of the forest belt seems to have been inhabited mainly by Bushmen, hunters and food gatherers, until the occupation of the area by the Bantu. The origin of the Bantu is doubtless still debatable, but Murdock (1959) quotes linguistic and other evidence which places their home of origin in a restricted area of the Cameroon highlands and a thin lowland strip connecting the latter with the coast opposite Fernando Po. Murdock considers that the Malaysian food plants such as bananas, yams and taros had entered Africa via the East African coast and spread across to West Africa, and that these provided the means for the Bantu to penetrate the tropical forest. They then worked across to the east, and there adopted the *Eleusine* millet and sorghum from the Cushite peoples. There is evidently room for argument about this reconstruction of Murdock's, but there can be no doubt that sorghum enabled the Bantu to expand into the savannah country of East Africa and southern Africa below the tropical belt, and that this expansion is of relatively recent date, having taken place mainly during the past thousand years. The fate of the Cushites is uncertain, but they and their agricultural practices are likely to have been absorbed by the Bantu. The appearance of many of the Bantu people in Tanganyika suggests that their origin could have been a West African stock intermarried with the Cushites, and the older view of the origin of the Bantu peoples was that they came from the intercrossing of Negroid and Hamitic peoples. The very name Tanganyika could mean mixture.

Burkill (1937) has pointed out the similarity between the East African and Central to South African sorghums, and we have already noticed that more than half of the varieties of southern Africa occur also in eastern Africa. There is no suggestion of any connexion between the West African sorghums north of the Gulf of Guinea and those south of the forest belt. The variety of *S. nigricans* which occurs in Nigeria and Rhodesia is also found in Tanganyika, Uganda and the Sudan.

Before considering further the variability in the sorghums, I will sketch briefly the spread of the crop from Africa. We know that cultivated sorghum had reached the Middle East by about 700 B.C., for a carving from Sennacharib's palace at Timgad depicts a sow with her litter feeding in a field of what is clearly sorghum (Piedallu, 1923). Herodotus, writing in 443 B.C., stated that the Babylonians grew millet, Kenchros, as tall as a tree (Tackholm, Tackholm and Drar, 1941), and this plant is likely to have been sorghum. It seems that sorghum reached the Middle East from India. The archaeological evidence indicates that it was not grown in Egypt before the Roman Byzantine era, so it did not travel to the Middle East through Egypt. A general Indian name of sorghum is jowar, or juar, and the Persian name is juar-i-hindi, suggesting an Indian origin for the Persian material. Further, Pliny records the movement of cultivated sorghum from India to Italy about the middle of the first century A.D. Sorghum must therefore have been in India for some years prior to the carving in Sennacharib's palace, so it was there about the beginning of the first millennium B.C. Sir George Watt (1893) pointed out that there was no specific Sanskrit name for sorghum, the most usual term being 'Yavana', which meant a stranger. It is therefore likely that cultivated sorghum reached India after the arrival of the Sanskrit peoples.

Assuming that the latter event may be dated during the first part of the second millennium B.C., we have a date for the arrival of sorghum in India between 1000 B.C. and 1500 B.C. It probably went in sailing vessels from the East African coast by way of Arabia and down to India, as the sea traffic on the monsoons began in the distant past. Certainly there was trade between the East African coast and the Arabian state of Ausan prior to 700 B.C. (Cole, 1954). Incidentally, the dhows from Arabia used to stock up with sorghum from the Tanganyika coast for ship's provisions until only a few years ago.

Sorghum reached the mainland of China at a fairly late date, probably not before the thirteenth century A.D. Burkill (1953) considers that it went across the Sabaean lane from Africa to China, but there are affinities with the Indian sorghums, and Ball's (1913) original suggestion that it went from India to China some ten or fifteen centuries ago still seems a probable guess. On the mainland of China a distinctive group of sorghums has arisen, the Kaoliangs. The sorghums of the Chinese coast, however, are the so-called Chinese Amber Canes, which are very different, and which Snowden classified under *S. dochna*. They are found in India and Burma, and Korea. Snowden classified a variety Collier under the same variety of *S. dochna* as these Amber Canes, and Collier is known to have been introduced from Natal to America in 1881. Its distribution suggests a movement by sea traffic, along the coast between Africa and China. This is very probable, since Chinese coins of the eighth century A.D. have been found at Kilwa (Coupland, 1938) on the East African coast, and a Chinese geographical work of the twelfth or thirteenth century A.D. refers to places on that coast.

Having briefly considered the movement of the cultivated sorghums, we will return to Africa to examine the relationships between the wild and cultivated types there. Anyone who has grown cultivated sorghums in areas where wild ones are also present must have noticed the frequency of natural hybrids between the two. They are to be found in the Sudan and East Africa, and also in the United States. In the Lake Province of Tanganyika such hybrids are termed 'pungu' in the local language. There are numerous specimens in the Kew herbarium that are certainly wild × cultivated crosses. Under these circumstances, we should expect a good deal of introgression, with gene flow both from the wild types to the cultivated, and in the opposite direction. The studies which I have made so far at Serere leave no room for doubt that such introgression is occurring between the local *S. verticilliflorum* wild type, and the cultivated forms. The cultivated sorghums usually show around 5 per cent of out-pollination, although this varies with weather conditions and panicle type. The wild Eusorghums probably outcross to a far greater extent judging from work done in the United States (Barton, 1951). When plant characters are scored and summed, the wild type shows a range at one end of the scale, and the cultivated type a range at the other. Intermediates occur, and in one study using scores of twelve plant characters a typical 'pungu' from the Lake Province of Tanganyika obtained the same intermediate score as another putative hybrid from Arua in the West Nile area of Uganda, some 500 miles away. The variability of the intermediates which persist for several seasons is in fact quite restricted, although when a cultivated × wild cross is grown on under good agricultural management a wide range of recombinations is obtained in the F_2.

Many of the sorghum growing people are aware of the influence of the wild sorghums on their cultivated types. In the Lake Province of Tanganyika, the grower removes hybrids from his fields as soon as they can be recognized by the deciduous sessile spikelets with hard black glumes. There is at Kew a pleasing note by a local African attached to a specimen of *S. verticilliflorum* collected by L. Thomas in the Rukwa area of Tanganyika. It reads: 'This plant some time ago was planted as corn by our people. They planted like maize for many times passed over, then the plants changed into grass'. This indicated an observed association between the wild sorghum and the cultivated types.

I recently obtained by chance a natural cross from a *S. verticilliflorum* collection, and studied the F_1, F_2 and F_3 generations. The plant originally collected was awnless, but did not lie outside the range of the wild species for any other character. Among its progeny, it produced nine plants which were tall with broad leaves, and were evidently an F_1, and thirty-two other plants which resembled the original collection. An F_2 was grown from the F_1 plants, and gave a good 3 to 1 ratio for deciduous and persistent sessile spikelets. Some of the F_2 plants had numerous cultivated characters, and those with persistent spikelets and large grains were grown on as an F_3. From this F_3 of over 3000 plants, fifty-seven plants were obtained which were identified and named by the local people as sorghum varieties in general cultivation for food or beer. A number of sterile plants also appeared, 70 per cent of the F_3 rows yielding one or more sterile plants. This was unexpected, as there had been no signs of poor germination or sterility in the previous generations. Thus, it is fairly certain that the wild and the cultivated varieties are influencing each other's characteristics. We need more facts, and have started a programme of reciprocal wild × cultivated crosses and back-crosses.

The situations in which the wild Eusorghums of Africa occur are of some interest. If we ignore the four African species based on single collections, we find that the remaining twelve all have habitats which could suit escaped weeds of cultivation. Some specimens of every one have been collected in or near cultivated lands, or on sites of old cultivation, excepting *S. vogelianum*. Of the nine specimens of the latter which I have seen, three are without precise locality, one was found opposite a railway station, and the remaining five were collected from river banks. Indeed, the edges of rivers, dry watercourses, and the margins of swamps are usual sites for these Eusorghums as well as cultivated land and land abandoned after cultivation.

Snowden (1936) has suggested that his species of cultivated sorghums are derived from wild species, principally *S. arundinaceum*, *S. verticilliflorum*, *S. aethiopicum* and *S. sudanense*. Their botanical characters are related, and if we plot the geographical area of the wild and cultivated Eusorghums, the association of each wild species with those sorghums which Snowden regards as derived from it, is very close. Gene flow between the wild and the cultivated Eusorghums, however, would have resulted in their mutual modification, and could account for these observed similarities.

The most satisfactory picture is that of a crop and its weed, man selecting the cultivated sorghum in certain directions, and a few of these modifications being transferred to the weed

Eusorghums, natural selection eliminating quite a lot. The weeds would escape from cultivation, and their progeny re-establish later in cultivated fields, thus tending to influence new sorghum types in the direction of those previously there. We would expect man to select in the high rainfall areas of West Africa, types of sorghum different from those in the arid areas of the Sudan or the moderate rainfall areas of East Africa, as in fact he has. Through introgression, we should expect the wild Eusorghums to reflect these differences, giving *arundinaceum* type in West Africa, *aethiopicum* type in the Sudan, and *verticilliflorum* type in East Africa. Incidentally, there is a good deal of variation within these wild species, especially in height, leaf length and width, spikelet size, and awn length.

The simplest picture would be that of cultivated sorghum spreading out from one original centre, with the weed form being carried along in the crop. It is not necessary to assume that wild Eusorghum was widespread in Africa. It could have been localized originally in the north-east quadrant. We have already noticed that the sites of the present day wild Eusorghums would suit an escaped weed of cultivation, and Snowden (1955) remarks on the type *S. verticilliflorum* that it may well have originated somewhere in the vicinity of the Great Lakes and been carried southwards along the lakes and rivers of the western and eastern Rift valleys by the Bantu races as they travelled south. This picture may be too simple. We cannot rule out the possibility that the wild Eusorghums were scattered through Africa prior to the spread of agricultural man with the cultivated crop.

Still thinking primarily of the position in Africa, we may now enquire how many species of *Sorghum* really are involved. Is the picture that of diversification within one large, polymorphic species? Or is the present situation that they are really two large polymorphic species, the wild and the cultivated? Or can both legitimately be broken down to numerous species deserving of binomial Latin nomenclature?

The real answer to these questions can only be found by growing very large collections of the wild and cultivated *Sorghums*, and analysing them and their hybrids. In my opinion, the slight evidence at present available does not warrant the erection of species within either the wild or the cultivated groups, in

Africa. All the cultivated sorghums are completely interfertile, and the only suggestion of differentiation of which I am aware is from the work of Kidd (1958), who in crosses between Kafir of South African origin, and Dura of Sudan origin, found that all the expected recombinations on the assumption that they are one and the same species do not occur. It looks as though one has a series of clines, a situation where there is incipient speciation. One can refer to a *caffrorum* from South Africa, a *guiniense* from West Africa or a *subglabrescens* from the Sudan, and the term conveys a definite meaning. However, the species boundaries are vague, and there are all sorts of intermediate forms. It seems best either to follow Hackel, who classified all these forms as varieties of one species, or else to split them as sub-species, varieties and cultivars. In America, the sorghums are usually grouped under *Sorghum vulgare* Pers., and then referred to by common names such as milo, kafir, and hegari. This is a very practical system, but the name *S. vulgare* Pers. is illegitimate, and the proper term is *S. bicolor* (Linn) Moench (Clayton, 1961). The case for such drastic lumping is that it retains the concept of a species which carries a Latin binomial, however difficult this may be to define. On the other hand, there is such a mass of material that it has to be classified somehow, and many find a grouping such as that of Snowden useful. We may consider most of Snowden's species as sub-species.

The situation regarding the wild Eusorghums of Africa is parallel, and almost certainly taxonomically similar, though we have less evidence. To me it seems best to classify all the wild ones under one species, which would probably be *S. arundinaceum*, split as sub-species, varieties and forms, but some would disagree strongly with this approach, and the matter can only be decided by a study of a large collection.

There is some reason to treat the wild and cultivated *Sorghums* as separate species. We have seen that only restricted intermediate forms survive in nature when the two are crossed, and the sterility of the F_3 generation of the hybrid reported earlier is also suggestive. Moreover, Bhatti (1959) found that his stocks of Kaoliang when crossed to *S. virgatum* (Tunis grass, one of the wild Arundinacea occurring in North Africa) produced seed which failed to germinate. Other cultivated sorghum stocks crossed with *S. virgatum* showed reduced pollen fertility in the F_1. The segregation ratios of various characters were disturbed, and there is little doubt from his results that *S. virgatum* is specifically distinct from the cultivated types he used. On the other hand, Sudan grass has been used extensively in breeding programmes with cultivated sorghums in the U.S.A. without reports of any sterility or other barriers, but it seems likely from the rather persistent pedicelled spikelets, and also from the spikelet shape, that Sudan grass has been much influenced by introgression from the cultivated forms.

Allowing then that there are two distinct species, we turn finally to the interesting question of how the differences between them arose. There are two possibilities. I was originally of the view that there were two species initially, with man ennobling one by selection. Later, the second species would have been picked up as a weed, as man moved the cultivated type into its area. From then on introgression between the two would have taken place, leading to the present situation. I visualized one species as a tall grass with broad leaves and large spikelets, and the second as a smaller grass some 3 ft tall with narrow leaves and smaller spikelets. This is certainly one possible view, but Professor Hutchinson drew my attention to an interesting alternative. It may be that we have in *Sorghum* a practical illustration of the process of disruptive selection which Professor Thoday has demonstrated in *Drosophila*. The two selection forces would have been man operating in the direction of cultivated characters, and natural selection operating to hold the species in a form suited both to survival in the wild and to persistence as a weed of cultivation. Thoday and Boam (1961) obtained a polymorphic population in *Drosophila* after thirty-six generations, with 50 per cent gene flow. With 25 per cent gene flow, a polymorphic population was obtained in twenty-five generations, and the divergence was greater. It is quite possible that the cultivated sorghum arose from the wild in this way, but there is as yet no critical evidence available. What is certain is that the conditions in Africa have been very suitable for diversification of the crop. Individual tribes have their own cultivated sorghums, usually several of them, which they maintain by selection. There has been a lot of movement of peoples, and there must have been periods of isolation, followed later by hybridization, when groups came in contact again. Constant selection has been practised, with an eye open for anything which might be of value—the African cultivator will never root up a food plant even when it is growing in a most unlikely situation—and intermittent hybridization has occurred with the wild Eusorghums. As yet, we know of no barriers to gene movement between any of the Arundinacea *Sorghums* of Africa, except that one cultivated group in Ghana is cleistogamous. We begin to appreciate why the systematist has such a formidable task.

We have noticed the interaction of wild and cultivated *Sorghums* in Africa. In India, the cultivated types came in contact only seldom with diploid wild *Sorghums* with which they were interfertile. We have noticed that crossing does occur with the tetraploids, but that frequently the progeny are themselves tetraploid. The tendency would therefore be for the cultivated crop to have a bigger influence on the wild Halepensia than vice-versa. There is a lot of diversity in the tetraploid Halepensia of India, some of which may well have arisen from this source. In South China, the cultivated crop will have made contact with *S. propinquum*. Mr Sieglinger of Oklahoma sent me some seed of various crosses between *S. propinquum* and kafir, and some of the segregates have many of the distinctive characteristics of the Kaoliangs. Again, we are guessing, but it may be that *S. propinquum* has played a part in shaping the Kaoliang group of sorghums in China.

To summarize, it looks as though cultivated sorghum was first developed in the north-east quadrant of Africa some 5000 years ago, probably in the Abyssinia-Sudan region. From there it spread to West Africa quite early in its history. It was carried into East Africa, and later to Central and South Africa. The cultivated and wild *Sorghums* have constantly influenced each other. Possibly two species were involved in the history of the present cultivated and wild types, or possibly both have developed from one wild species through disruptive selection, followed by a history of isolation, hybridization and selection in numerous permutations and combinations. From Africa, the cultivated crop moved to India at the end of the second millennium B.C., or early in the first millennium. It spread from there to the Middle East, having reached Syria by 700 B.C., and to China perhaps only about 1000 years ago. In India, there may have been some introgression between the cultivated crop and the tetraploid wild types, mainly from the former to the latter. In China, there has probably been introgression with *S. propinquum*.

This lecture has really been given ten years too early. It is only possible now to present an outline of the probable history of the crop, including much theory and few facts. We can hope to collect much more information over the next ten years.

IV

THE COMPARATIVE PHYLOGENY OF THE TEMPERATE CEREALS

by G. D. H. BELL

THE origin of the cereals of the north temperate regions, and their evolution and development in cultivation, are of the utmost interest and significance because of the intimate association of botanical evolution with the sociological development of man and the processes of civilization over a considerable area of the earth's surface. It is generally recognized that the domestication of the two 'primary' cereals, wheat and barley, was an essential feature of the oldest civilizations that are known, and the ultimate domination of man, and his colonization of the new continents, have been made possible by these two grain crops, but supplemented by the two 'secondary' crops, oats and rye. The four cereals have acted in an extraordinarily complementary way in providing the needs of temperate agriculture, and thereby furnishing the requirements of civilized man and his livestock.

The agricultural history and present status of these four crops are, of course, reflections of their biological evolution with particular reference to the emergence of cultivated forms, and the establishment of the level and range of genetic variation, which characterize each. Taxonomic structure, phylogenetic relationships, reproductive mechanisms, genetic architecture, and all the important biological features of a crop determine not only the methods and techniques of breeding, but set the limits to improvement by genetic manipulation. The history of the more recent work of plant breeders in the cereals reflects very accurately the exploitation of what each has to offer in terms of these biological concepts.

It is impossible to consider the phylogeny of any group of plants without some understanding of its systematic and taxonomic botany. In terms of classical taxonomy, the classification of cultivated plants of great antiquity offers special difficulties usually not met with in dealing with the phyla and taxa of wild plants, and concepts of genera and species often need adjustment if reasonably sensible decisions are to be made. These domesticated plant groups are special cases, otherwise they would not have reached their important position as cultivated plants, and some of the cultivated cereals are eminently suitable for certain fields of investigation, particularly genetics and cytology, with the application of new concepts and ideas to the study of taxonomy and phylogeny. In particular, there has been a unique contribution from the study of cultivated wheat and related groups to an understanding of the significance of polyploidy, genome structure and individuality, and the concept of chromosome homology, all of which are pertinent to a study of phylogeny and to the development of natural systems of classification.

Obviously the characteristic features of the cultivated cereals derive from the family—the Gramineae—to which they belong, and the most up-to-date appraisal of the Gramineae has been more strongly influenced by recent ideas on certain fundamental taxonomic concepts than have most families. This is undoubtedly due to the amenability to some modern methods, particularly cytogenetics, of taxonomic investigation of some important groups in the family. Therefore, in spite of certain taxonomic difficulties inherent in the Gramineae, considerable advances have been made from detailed studies in the family regarding phylogenetic relationships.

The four temperate cereals emphasize and illustrate these points very clearly, as well as bringing out into strong relief certain dominating considerations when discussing evolutionary processes. Most significant of all is probably the role of polyploidy. The Gramineae as a family is outstanding in its high incidence of polyploid forms, Stebbins (1956*b*) calculating that nearly 70 per cent of grass species are polyploid, a figure which is double the average of flowering plants as a whole. This incidence of polyploidy is associated with hybridization so that there has been a striking development of allopolyploid groups at different levels of ploidy, and the four genera with which this discussion is concerned each offers a characteristic picture with regard to the involvement of allopolyploidy in the evolution of the cultivated forms.

The basis of any phylogenetic ideas when considering the cultivated plants is, of course, the identification of the wild prototype or the nearest wild progenitor or progenitors and by what evolutionary processes the range of known genetic variation has arisen. In the cereals it is generally assumed that the progenitors, or prototypes, were grass-like forms, and that they would possess certain wild, or primitive, characters. Failing the direct and accurate identification of any prototype, it is customary to build up a picture of what is regarded as primitive, the characters associated with primitiveness, and the unfolding and evolution of cultivable forms. There is, in the cereals, one very important common denominator, namely that a sequence is perceptible from forms possessing primitive characters, or 'wild' characters, usually associated with seed dispersal or dissemination, to the specialized characters required for cultivation. These, in general terms, are the inflorescence or spikelet characters of rachis fragility, spikelet articulation, awn development, basal spikelet soil-boring modifications, hairiness, and small grain size. There is in some cases a transition from the true wild type to the weed type, which occupies cleared ground or cultivated land, and the wild progenitor, or the putative ancestor, must necessarily vary in its relative primitiveness depending on how far back in evolution it has been possible to trace indisputable relationship.

Whatever else may be taxonomically contentious and phylogenetically doubtful in considering these cereals, it seems to be generally permissible and agreeable to distinguish and separately name four distinct genera—*Triticum*, *Secale*, *Hordeum* and *Avena*—as the taxonomic units which include the cultivated wheats, ryes, barleys and oats, as well as their most closely related non-cultivated forms. It is also reasonable to assume that *Triticum*, *Secale* and *Hordeum* had a common origin, and they are naturally associated in the tribe Triticeae (= Hordeae) which includes other important and interesting genera such as *Aegilops* and *Agropyron*. The concept and identity of the Triticeae with its sub-tribe Triticinae, in which are included *Triticum*, *Secale* and *Aegilops*, and the sub-tribe Hordeinae, in which are the genera *Hordeum*, *Elymus* and others, are fundamental to a proper understanding of the taxonomy and comparative phylogeny of wheat, rye and barley. *Avena*, on the other hand, is remote from the Triticeae, being the type genus of the Avenae, a tribe which has not contributed any cultivated cereals other than oats.

Wheat

It is a remarkable fact that the Triticeae has given rise to the cultivated wheats, barleys and ryes, thereby contributing three of the four temperate cereals under discussion. The close phylogenetic relationships of the genera usually recognized in this tribe are apparent in the high number of bigeneric hybrids that are known, or have been produced artificially, a position which is said to be unique among higher plants. However, this very circumstance has resulted in considerable doubts as to whether it is taxonomically correct, and phylogenetically realistic, to continue recognizing certain groups as belonging to distinct genera. Stebbins (1956*a*, *b*), for example, from a consideration of all the information, including that provided by chromosome behaviour and cytogenetic data, has concluded that the tribe is better envisaged as one genus, although it is admitted that there is not adequate data for devising a satisfactory taxonomic treatment on this basis.

This approach to the modification of the taxonomy of this closely related group, has been further emphasized by Bowden's recent proposal for dealing with *Triticum* and *Aegilops* within the sub-tribe Triticinae. Bowden (1959) considers that, based on the most recent phylogenetic interpretation of the cytogenetic data, as well as on Anderson's method of morphological extrapolated correlates, *Triticum* and *Aegilops* should be considered as one genus. The significance of this is that, in considering the evolution of the polyploid series in wheat in view of the recently established knowledge, the two closely related genetic systems hitherto recognized as of generic status, would be brought within the concept of a single genus. Thus greater emphasis is placed on the common origin of *Triticum* and *Aegilops*, and recognition is given of the modern interpretation of the evolution of the cultivated wheats. (Fig. 17.) This has only been possible by the determination of the origin of the tetraploid and hexaploid allopolyploid species by identifying the *B* genome as having come from *Aegilops speltoides*, and the *D* genome from *Aegilops squarrosa*. Thus, with the basic seven-chromosome *A* genome of the diploid wheats, there has been a synthesis of the three chromosome races *AA*, *AABB*, and *AABBDD*, with, in addition, a fourth genome *GG* which is probably a modified *BB* (Fig. 18).

The significance of this phylogenetic development in wheat is that both infusions of the wild goat grass (*Aegilops*) genotypes

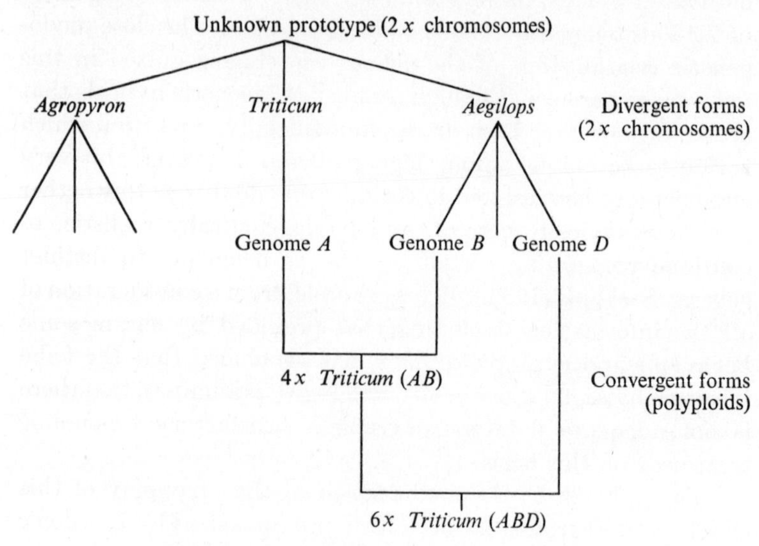

Fig. 17. Origin of polyploid species in *Triticum*.

have been responsible for initiating a new and wider range of genetic variation. The diploid *Triticum* forms exist in the wild, as typified by the *aegilopoides* forms, and also are represented by the basic cultivated *monococcum* series, which in many ways parallels in its range of botanical forms that found in *aegilopoides*, but has lost certain wild characters of the spikelet, though retaining the enclosed grain and the brittle rachis (Plate V). There is sufficient variation in the wild diploids to account for the variation known in *monococcum*, the primitive einkorn which is known from the earliest civilizations and persists in isolated areas in Europe today. Helbaek (1960*b*) has recorded *mono-*

coccum at Jarmo, which is the earliest agricultural site known showing evidence of plant husbandry and is estimated to have existed as early as 6750 B.C.; but significantly Helbaek concludes that einkorn played only a subsidiary role in the early development of cultivated wheat.

It is apparent that the cultivated wheats would not have progressed far at diploid level, and it is interesting that the basic tetraploids, *dicoccoides* and *dicoccum* parallel very closely the *aegilopoides* and *monococcum* series, but with a more robust

Triticum

		Diploids (2*x*)	Tetraploids (4*x*)	Hexaploids (6*x*)
Wild	Grain invested	*aegilopoides A* *thaoudar A*	*dicoccoides AB* *armeniacum AG*	
	Grain free			
Cultivated	Grain invested	*monococcum A*	*dicoccum AB* *timopheevi AG*	*spelta ABD* *speltiforme ABD* *rigidum ABD* *macha ABD* *vavilovi ABD* *tubalicum ABD*
	Grain free		*durum AB* *turgidum AB* *polonicum AB* *pyramidale AB* *orientale AB* *persicum AB*	*compactum ABD* *sphaerococcum ABD* *vulgare ABD*

Fig. 18. Wheat species and their genomes.

morphology (Plate VI). These basic tetraploids are represented at Jarmo, and Helbaek refers to transitionary forms between the wild *dicoccoides* and the cultivated *dicoccum*, or emmer, types. These show a wide variability in contrast to the more refined and uniform type characteristic of the neolithic emmers which show the obvious effects of selection by man, and became the dominant cultivated wheat of early civilizations, spreading to many important centres in Europe and persisting in cultivation until the present day. It is interesting, however, that archaeology has not contributed anything of significance as yet to the history of the free-threshing tetraploids before their occurrence in Egypt and in Greco-Roman times, and there is no evidence that such important forms as *durum* (macaroni wheat) or *turgidum* (Rivet wheat) occurred in prehistoric association with the other wheats. According to Helbaek, it is the hexaploid club wheats (*compactum*) that temporarily supplanted emmer in some areas, such as Mesopotamia and Egypt, as early as 3000 B.C. and consequently they must have arisen at a very early date in the history of agriculture.

Although the major steps in the establishment of the tetraploid and hexaploid allopolyploid species of *Triticum* are known in so far as the origin of the alien genomes has been identified, there is still much that is a matter of surmise in visualizing the exact course of events. It is, therefore, worth while considering in more detail the available evidence regarding the sequence of differentiation from the wild and cultivated diploids, through the tetraploids with the wild and cultivated enclosed-grain forms to the free-threshing types with tough rachis, and culminating in the hexaploids with cultivated forms only, dominated by free-threshing forms, but with the enclosed grain and special brittle rachis persisting in the *spelta* forms. Hexaploidy, in terms of the *aestivum* species or group, is generally considered to have conferred on cultivated wheat its greatest potential, and it is certainly true that at this level wheat has achieved its eminence as a world crop. It would be logical to conclude that *monococcum* differentiated from *aegilopoides* or some similar form, and it is interesting that *monococcum* is considered by some workers to be the key form at diploid level and the donor of the *A* genome to the tetraploids because it appears to be closest to the Sitopsis section of *Aegilops* which again is considered to be transitional to *Triticum*. However, circumstantially it might be expected that *Aeg. speltoides* hybridized with an *aegilopoides* type to produce the wild tetraploid similar to *dicoccoides*. The distribution of the species allows this. Sarkar and Stebbins (1956), however, refer to *monococcum* and its relatives as the wheat parent, which might imply an origin for the tetraploids under cultivation. The synthetic amphiploids produced by hybridizing both *aegilopoides* and *monococcum* with *speltoides* really give no clue to the correct answer, although Riley and others (1958) state that the original tetraploid that arose in nature would have been very similar to the *monococcum* × *speltoides* amphiploid that they produced.

The morphological approach, and the study of karyotype and chromosome behaviour have gone a long way to showing the origin of the tetraploid wheats and the synthesis of the *AB* genomes. It has been shown, of course, by Riley (1960) that a gene in the *B* genome (chromosome V) controls meiotic pairing and hence the tetraploid settled down to a diploid-like behaviour with perfect bivalent formation restricted to completely homologous chromosomes within each genome. It might also be stated here that the tetraploid *timopheevi* with genomic structure *AG* behaves very like *AB* tetraploids when hybridized with *speltoides*, indicating a similar origin for the *B* and the *G* genomes as suggested originally by Sachs (1953).

The basic tetraploid wheat type can be visualized as a *dicoccoides*-like form, and presumably giving rise directly to *dicoccum* as the first cultivated tetraploid. But where do we turn for the next stage of development? The cytological, genetic and morphological evidence would suggest that the cultivated tetraploids could be derived from the *dicoccoides* wild type, and there is little difficulty in ascribing this origin to the species such as *durum* and *turgidum* among the free-threshing forms. *Orientale* has been regarded as a primitive free-threshing tetraploid, while *persicum* (= *carthlicum*) has been considered as the link between the tetraploids and the hexaploids. Geographically the tetraploids have their centre of diversity, and probably of origin, in the Mediterranean basin, whereas the centre of the diploids was Asia Minor.

As far as species differentiation is concerned, genetic analysis has indicated quite clearly that certain of the basic characters, such as glume shape, fragility of the rachis and solidness of the stem etc., are identifiable by individual gene action referable to individual genomes. Thus in the tetraploids, there is the outstanding example of the K, K_3 pair, in which K confers the loose glume and tough rachis to *durum* and *turgidum*, while K_3 is responsible for the enclosed grain and brittle rachis of *dicoccum* (Watkins 1930). A third allele, k, determines the *persicum* glume and rachis type. These three alleles differentiate species also at hexaploid level, modified by K^d in the D genome, and are therefore responsible for the major specific differences at tetraploid and hexaploid levels. Other examples of specific genes are well known, such as those that differentiate the hexaploid *sphaerococcum*, *compactum*, *aestivum*, *macha*, and *vavilovi*.

Mac Key's (1954*a*) work with artificially induced mutations in hexaploid wheats has undoubtedly helped to elucidate speciation at hexaploid level, and gives valuable evidence of phylogenetic relationships. He proposed a *Q* factor which he identifies with Watkins' *K* factor, and also with Philipschenko's squarehead gene *q*, and he envisages the *Q* factor as interfering with the action of several genes controlling major morphological characters. This so-called suppressor factor, *Q*, is known to be situated on the long arm of chromosome 5A in the *A* genome, and apparently must have arisen as a mutation in a tetraploid wheat with enclosed grain and brittle rachis. As Swaminathan and Rao (1961) point out this suppressor factor, with the multivalent suppressor factor on the long arm of chromosome 5B of the *B* genome, has had outstandingly important effects on the tetraploid and hexaploid wheats, while the differentiation of *compactum*, *sphaerococcum* and *spelta* from *aestivum* by single genes respectively shows the vital significance of macro-mutations in the evolutionary history of wheat. Swaminathan and Rao envisage the hexaploid wheats as existing in two basic morphological types: first, *aestivum sensu stricto* from which *compactum* and *sphaerococcum* differentiated; and *spelta* from which *vavilovi*, *macha* and *zhukovskyi* differentiated. This concept would support the view that all hexaploid wheats can be included as subspecies of *T. aestivum* L. and certainly strengthens the idea that macro-mutations are the basis of the genetic and morphological differentiation of polyploid wheats.

Although it has been demonstrated clearly that the origin of the *D* genome of the hexaploids is from *Aeg. squarrosa*, while the *A* and *B* genomes of this group are homologous with the *A* and *B* of the tetraploids and the synthesis of *spelta* indicates an origin from *dicoccoides*, there is still a great uncertainty of the unfolding story in the hexaploids (Plate VII). McFadden and Sears (1946) consider that *spelta* is the primitive hexaploid type, and is the progenitor of the free-threshing European hexaploids, *aestivum* and *compactum*, and that the Asiatic hexaploids such as *speltiforme*, and *macha* had a similar origin to *spelta*. If this is so, then these authors conclude that *aestivum* and *compactum* are of comparatively recent origin. Mac Key, however, does not subscribe to this view, although he agrees that the original hexaploid wheat 'had a more or less fragile rachis'. His view is that *compactum* may have been the initial form of a hexaploid with a tough rachis, that *compactum* preceded *aestivum* in cultivation, and that *spelta* arose from the hybridization of *compactum* (*antiquorum*) with *dicoccum*. There is, of course, the possibility that the ancient free-threshing wheat of Europe—the Lake Dweller wheat—cultivated in Neolithic times, was a dense ear tetraploid, from which *persicum* was derived. This would dispose of the supposed antiquity of any hexaploid free-threshing form, and give McFadden and Sears support for suggesting that the all-important *aestivum* arose from the hybridization of *spelta* and the Lake Dweller wheat. More recently Kihara has suggested that *aestivum* arose from *spelta* × *compactum*, both of which were derived from *macha* by mutation.

The really significant fact about the hexaploid cultivated forms is, however, that of the three species one (*sphaerococcum*) is practically monotypic and confined to North India, another (*compactum*) is less restricted in variation and adaptation, while the third (*aestivum*) is the most successful of all the wheat species and occupies the great bulk of the world's wheat crop. It is difficult to see how *aestivum* can be a young species. It is highly differentiated morphologically, physiologically and ecologically, and what perhaps is even more significant, it does appear to have chromosome differentiation resulting in, for example, a distinction between geographical groups such as the Western European as opposed to the Asiatic. These are karyotypic characters which would indicate a long history. Is an *aestivum* type the progenitor of the free-threshing hexaploids? In many ways this is more probable than the primitiveness of *spelta*, which has many features of a derived form. Riley (private communication) considers that there is a species or sub-species complex of *macha*, *compactum*, *spelta* and *aestivum* with no certain evidence of evolutionary sequence (Plate VIII).

The amenability of wheat to the type of detailed genomic analysis to which it has been subjected strongly suggests, if not implies, that the genomes of tetraploid and hexaploid wheats are, as Riley has pointed out, 'in essentially the same state as those of the original diploids from which they were derived'. This means there have been no structural chromosome changes of any size or importance. In spite of this there is no pairing between the different genomes in hexaploid wheat, and this means that the affinity between equivalent chromosomes of the different genomes must have been suppressed or removed. With the establishment of the seven groups, each of three pairs in the hexaploid, it has been possible to recognize seven homoeologous groups, each of three chromosomes, and each of these homoeologous groups has one chromosome pair in each of the three genomes. Homoeologous chromosomes may be said to be genetically equivalent, occurring in the three genomes and derived from the original diploids.

The cytological organization of the hexaploid wheats, as demonstrated by Sears (1948), by Riley and others, is considered to have greater evolutionary significance than the organization in those polyploids with polysomic inheritance. The disomic inheritance found in certain polyploids does allow alleles at duplicated loci to mutate without deleterious effects, because the original function is continued at homoeological level. Undoubtedly the development of this diploidizing mechanism has been of the greatest significance in the evolution not only of the hexaploids in wheat, but also of the tetraploids. It is significant that the tetraploid *durum* is second only to the hexaploid *aestivum* in its contribution to agriculture by providing a wide range of valuable cultivated forms. Indeed, the virtues of *durum* have prompted Shebeski (1958) to question the value of the addition of the *D* genome as far as its contribution to agricultural types is concerned. This certainly raises the question as to whether more valuable forms can be synthesized by the artificial production of amphiploids, as was suggested by Bell and others (1955), although current work would indicate that alien chromosome substitution or addition is likely to be more profitable from the breeding point of view.

The great range of genetic variation of *aestivum* is the result of a considerable period of evolutionary development, and it would hardly be hoped that any raw polyploid could emulate it. The *D* genome certainly contributed an enormous potential, as well as being responsible for the unique feature of the bread wheats in terms of endosperm biochemistry. Whatever may be the comparative genetic simplicity in determining morphological differentiation at species level in the polyploid wheats, there is a much more significant and complex differentiation interspecifically, and intraspecifically in certain cases, which determines agricultural and economic value. There is no doubt that the phylogeny and cytogenetic structure of wheat offer greater possibilities for genetic reconstruction than do the other cereals now that the barriers to exploiting alien genetic variation are better understood and the means of overcoming them are apparent (Riley and Bell, 1958).

BARLEY

It is probably most striking to consider barley after wheat because of some extraordinarily contrasting phylogenetic features of these our most ancient cereals. As to which of the two was first domesticated is not known, but it is quite clear that barley was a most significant crop in the origin and development of early civilization. There is considerable archaeological, cultural and historical evidence for this, and ancient barley remains occur with emmer wheat and a kind of millet, and are referable to about 5000 B.C. This is worth mentioning at the outset as most of these ancient barleys in the Asia Minor and Mediterranean areas are of the cultivated type known as six-row, because of the number of rows of grain on the ear, in contrast to the two-row forms which, archaeologically, seemed to be more recent. In fact, according to some accounts, the most ancient record of a two-row barley is 300 B.C. in Greek and Roman archives, and this type of barley is said to have been rare in Europe until the sixteenth century, when it suddenly became common. But identification of these barley types is by no means certain from archaeological remains.

The significance of the distinction between two-row and six-row barley is that the main argument concerning the origin and phylogeny of cultivated forms is whether the primitive wild progenitor was a six-row or a two-row form. It should be mentioned that the genus is characterized by being heterospiculate, the only forms which are not being the cultivated six-row varieties, and the nearest wild six-row species called *agriocrithon*, which are isospiculate. Bowden (1959) considers that the ancestral stock of the genus was derived from an isospiculate ancestor, but the genus is basically heterospiculate and the isospiculate condition must be regarded as derived. No one has demonstrated with certainty which is the more probable, and much depends on views as to evolutionary trends in the Triticeae.

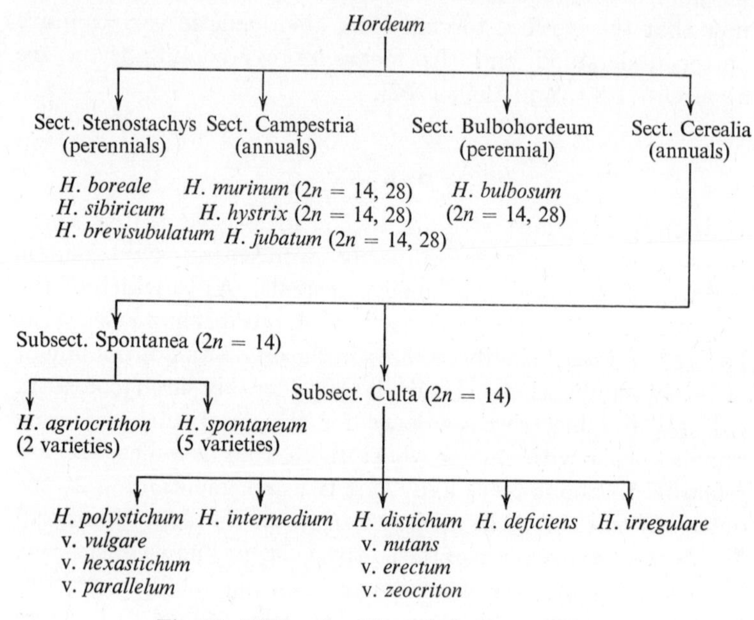

Fig. 19. Wild and cultivated *Hordeum* species.

There are something like twenty-five species in the genus *Hordeum*, some of which are wild and comparatively useless grasses included in three separate sections, while there is a fourth section—the Cerealia—which comprises the grain producers both 'wild' and cultivated. This exemplifies an interesting taxonomic contrast with wheat. The grass species may be diploid ($2n = 14$) and tetraploid ($2n = 28$), or diploid, tetraploid and hexaploid ($2n = 42$), or hexaploid only, but no polyploidy is known in the cultivated forms, or in the 'wild' grain types (Fig. 19). The grasses of this genus are genetically isolated from the grain producers in spite of certain similarities of some of the species with the grain types. No natural hybrids are known between the grain producers and the grass species, and artificial hybrids are difficult to produce, resulting in sterile plants when successful. Again, natural intergeneric hybrids, involving such genera as *Agropyron* and *Sitanion*, are with the grass species. The conclusion is, therefore, that phylogenetically the known barley grasses have no relevance to the origin of the grain-producing barleys, and it is not known how these latter arose. On the other hand, Oinuma (1952) has found that some wild diploid and tetraploid species—*murinum* and *nodosum*—have karyotypes of the a_2 and a_3 forms which are similar to those in cultivated barleys.

However, the origin of the Cerealia is not known. It must have been confined within the genus as known and recognized, but there has been a very long history of separation and divergence from the grass types some form of which must have been the prototype. In spite of the superficially promising position for being able to trace evolution from grass to cultivated cereal within this genus, one is forced to start at the stage of the 'wild' grain producer, as for example was the position in wheat with *aegilopoides*. But with barley there is no outside influence, and there is no suggestion of the type of synthesis seen in wheat which involved at least pseudo-intergeneric hybridization, and the building up of polyploid series with the recognition of genomic syntheses.

Cultivated barleys comprise a homogeneous group of inter-fertile botanical forms, differentiated by a surprisingly large number of discrete morphological characters of the ear. Something like 180 botanical forms have been recognized, while geographical and ecological races are recognized with a very large number of agricultural varieties. Barley has been highly successful in its evolution in spite of its closed origin and its persistence as a diploid. The great difficulty with cultivated barley has always been how to classify all the forms, and even today there is argument as to whether they should all be included in one species, or whether the major types, based on the fertility of the lateral spikelets of the triplets at each node, should be given specific rank. In other words there is no agreement as to what are primitive forms of cultivated barley, or what characters in themselves may be regarded as primitive. The argument, as previously mentioned, is virtually confined to whether the two-row or six-row condition is the more primitive. Bowden, perhaps as would be expected in view of his treatment of the wheats, has suggested that the section Cerealia should be included as one species (*H. vulgare*), with two subspecies which are essentially the wild types, while he includes all the cultivated forms under the heading 'groups of cultivars'.

The basic wild grain-producing form known for many years is *H. spontaneum* (Plate IX) and its variants—*ischnatherium*, *bactrianium*, *turcomanicum* and *transcaspicum*—which occur in northern Afghanistan, Iran, Iraq, Asia Minor, etc., and are regarded as truly spontaneous. They have all the desired characters of the wild type with the disseminating mechanism and means of inserting in the soil—called trypanocarpy by Zohary—well developed and equivalent to that seen in the wild forms of other genera. These were the only wild forms known until the discovery in 1938 of the brittle rachis six-row form known as *agriocrithon* (with its varieties *paradoxon* and *dawoense*) in Tibetan regions (Plate IX). A similar type was found in 1944 in two centres in Israel by Kamm, 1954, and *agriocrithon* was accepted by many authorities as the long sought primitive wild six-row progenitor of cultivated barley. Åberg (1940, 1957), on the basis of his study of *agriocrithon*, has put forward alternative views for the origin of cultivated barley (Fig. 20). The simple question is—which is more primitive—*agriocrithon* or *spontaneum*? Six-row or two-row? Zohary (1960) maintains *agriocrithon* is a derived form, and stoutly supports a monophyletic origin from his study of hybrid swarms of brittle rachis two-row and six-row forms in Israel. He is convinced of the great significance of the very specialized seed dispersal and burial adaptation of *spontaneum* in the evolution of *Hordeum*, and maintains that this specialization is very ancient and was developed at the beginning of the differentiation of the genus. If this is so, primitive agriculture would have had only two-row barley, and six-row forms arose only subsequent to the origin of the tough rachis two-row forms under domestication.

There is much to support Zohary, including particularly the limited distribution of *agriocrithon* and the fact that it has only been found virtually as a weed of cultivated crops, as have all the Asiatic primitive and linking forms so far described. But this does not mean that the ancient primitive wild barley was not a six-row (or *hexastichon*) type. Indeed Zohary agrees to this, and so does Takahashi (1955), but the form has yet to be found, and Zohary considers it is likely to be a perennial. It is suggested, however, that the earliest agriculturists did not know the primitive six-row wild type, but the Cerealia was established from a *spontaneum*-like type before settled agriculture. This presup-

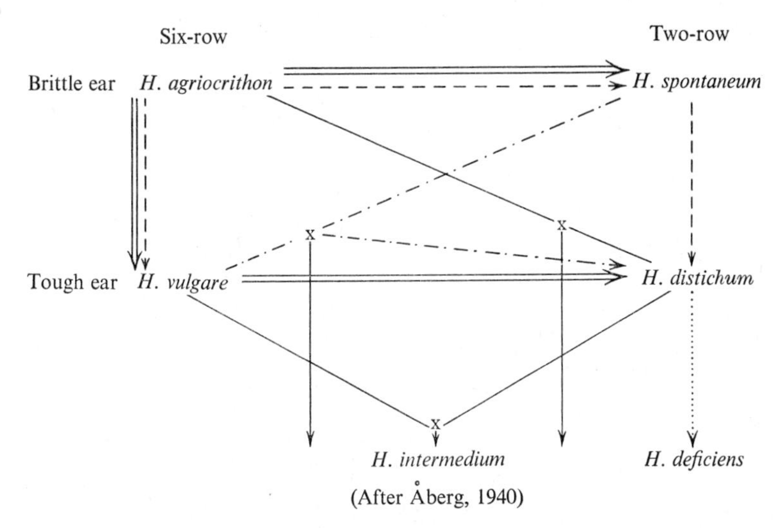

Fig. 20. Possible origins of cultivated barleys.

poses the reappearance of the six-row type under domestication after the tough rachis two-row forms had appeared, and the rigorous selection pressure for trypanocarpy had been relaxed.

Helbaek's recent finds of two-row non-brittle forms of barley at Jarmo in Iraqi Kurdistan, dated at the seventh millennium B.C., would give archaeological support to these views because they do suggest a link between *spontaneum* and cultivated barley of recent times. Support is also given to the classical ideas of the western Asian origin of cultivated barley, coinciding as it does with the distribution of *spontaneum*, with the eastern Asian and Ethiopian centres being secondary and resulting from introduction of cultivated barley.

It seems probable that the Jarmo barleys indicate a centre where differentiation by introgressive hybridization was in progress, as has been demonstrated as still taking place in Israel by Kamm (1954). Three centres in Israel showed a spontaneous active development of a wide range of botanical forms by hybridization of *spontaneum*, *agriocrithon* and cultivated forms.

Consideration may now be given to the available genetic

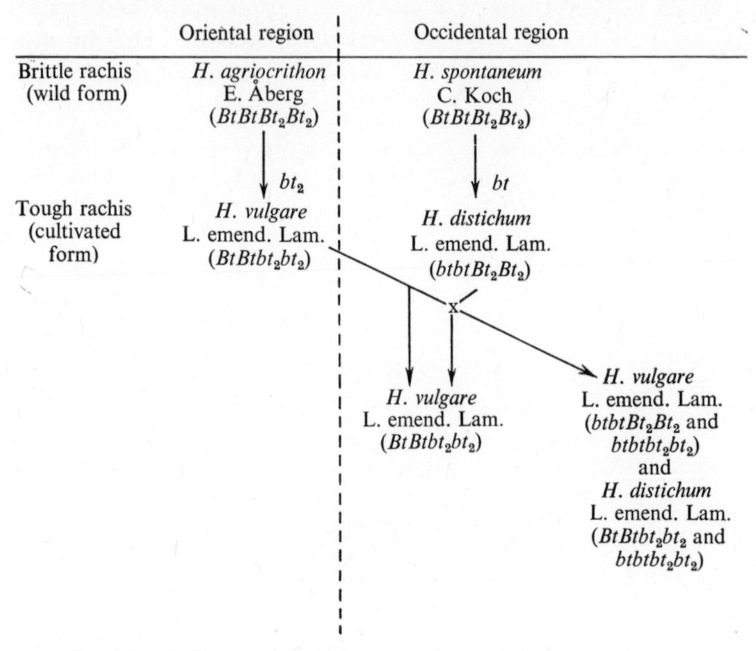

Fig. 21. Phylogeny and geographic differentiation in cultivated barleys (after Takahashi).

knowledge for the most acceptable phylogenetic views. As pointed out by Ubisch (ref. Takahashi, 1955), the wild and cultivated groups of barley are closely associated morphologically, and they share eighteen distinct alleles. The problem of the origin of the cultivated types, however, revolves around the two genetic systems controlling respectively tough and brittle rachis and the fertility of the lateral spikelets. Takahashi has shown that there is a difference in the distribution of the two genes controlling rachis toughness as shown by a survey of cultivated varieties (Fig. 21). *H. spontaneum* has the formula $BtBtBt_2Bt_2$, while western European varieties tend to be of the 'W' type—$btbtBt_2Bt_2$—and the eastern Asiatic forms of the 'E' type—$BtBtbt_2bt_2$. Mutation of the Bt_2 gene to bt_2 in East Asia would give cultivated six-row from *agriocrithon*, and of *Bt* to *bt* in South-west Asia of cultivated two-row from *spontaneum* on Takahashi's views, but Zohary considers both mutations occurred from *spontaneum* in West Asia, and the bt_2 gene was taken to Asia in a cultivated six-row barley. The historical sequence is, however, not known nor whether the mutant forms moved from east to west or west to east. It should be mentioned that Oinuma's karyotype survey of wild and cultivated barleys has shown that the longest of the seven chromosomes—chromosome *a*—gave five types a_1, a_2, a_3, a_4, a_5, depending on the length

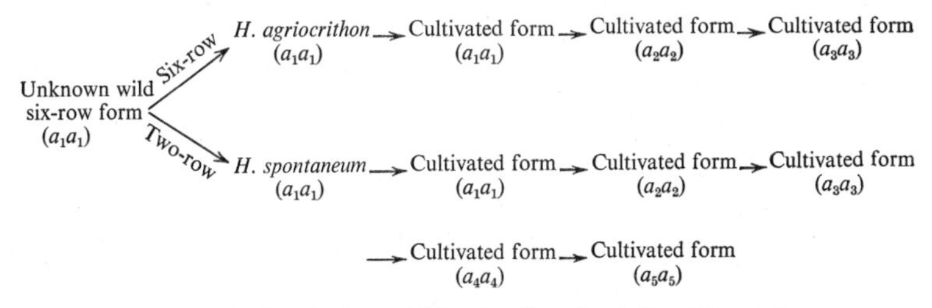

Fig. 22. Evolution within the Cerealia (after Oinuma).

and position of the secondary constriction. Both *spontaneum* and *agriocrithon* were of a_1a_1 type, as were most cultivated varieties, but a_2a_2 and a_3a_3 were also widely distributed, and Oinuma envisages a parallel differentiation from the two prototypes (Fig. 22).

But there still remains the problem of the primitiveness of the hexastichous or distichous conditions. Six-row is recessive to two-row and is dependent on a major gene difference, and mutations are known and can be induced from dominant to recessive. Zohary considers these would not persist in the wild. Dominant mutants are also known to be possible. The general fertility of the lateral spikelet is however really dependent on a series of multiple alleles, and changes are possible in either direction. Six-rowedness does appear logically the primitive condition, rudimentation is more probable than the reverse, and the extreme form of lateral sterility (the *deficiens* condition) is found in a restricted range of forms, is not known in the wild, and would appear to be of more recent origin. The whole matter would be cleared up, of course, if a decision could be reached regarding the alternatives put forward by Åberg, or if a truly wild six-row form could be found with all primitive characters expressed.

In the meantime, one has to accept the uncommon richness of this diploid group of cultivated plants in easily distinguishable and discrete genetic characters inherited in a comparatively simple manner (Plate X). With something like 200 variants recognizable and eighty-two cultivated, there is, however, a dominance of ten only, with three preponderating—*pallidum*, *nutans* and *medicum*. These possess many of the characters that would be associated with the truly wild type, but with the elimination of the primitive seed dispersal mechanism. It is interesting how cultivation and utilization of the barley crop has caused the survival and wide spread of the pigmentless, hulled, awned forms, while the special modifications of naked grain, hooded spikelets, and various expressions of colour in grains and hulls, have persisted only in restricted areas and often associated with primitive agriculture. Today, largely as the result of the work of Takahashi and his colleagues, it can be demonstrated that there is a definite association of genotype with region, and there is as previously mentioned a regularity in the occurrence of certain genes, some of which, the 'normal' characters according to Takahashi, are common to the wild progenitors, and others of which have arisen in cultivation and are special to particular areas. These latter are obvious in certain cases such as the hooded type of the Orient, the semi-brachytic type of Japan and Korea, the black, naked, hooded *deficiens* types of Abyssinia, etc.

The development, by plant breeding, of improved genotypes in barley must continue to be confined to the Cerealia section of *Hordeum*, and the considerable range of genetic variation in terms of morphological and physiological characters, and chemical characters of the grain exploited by hybridization within appropriate groups of the cultivated forms (Smith, 1951; Bell and Lupton, 1962). *Spontaneum* forms have been used in breeding programmes, and currently resistance to powdery mildew from this source is proving of value, but there does not appear to be any opportunity of exploiting disease resistance from more taxonomically remote sources as is being done in wheat. Most breeding programmes are, consequently, based on a comparatively restricted genotypic basis between very similar phenotypic types, although considerable advances have been made by using parental forms of this kind which derive from distinct geographical areas, and are themselves the result of intensive breeding programmes. Barley has great genetic resources, but obviously does not provide the scope in breeding techniques that the recent work in cytogenetics is offering for wheat.

OATS

Oats, which is usually referred to as a secondary crop, is of more recent origin in cultivation, as it is usually accepted that there is no real evidence that this cereal was known or cultivated by the Egyptians or the early Eastern civilizations. Indeed, prior to the Christian era oats were referred to only as a weed, and subsequently there is reference to its cultivation as a grain crop in central and western Europe, and for fodder in Asia Minor. The spread of oats is commonly thought to have depended on the spread of wheat and barley, the weed forms occurring in these crops and being taken into cultivation with the development of the cultivated characters. It is, indeed, claimed that oats, when it occurs in the wild state, is always adventive.

In these circumstances it might be assumed that the evolution of cultivated oats, and the relationships with 'wild progenitors', would be obvious, particularly as there are true wild forms and weed forms of agriculture, while the range of genetic variation and the basic structure of the genus are well known. The genus *Avena*, on modern concepts, is also well defined in relation to the other genera of the tribe Avenae, and is now regarded as comprising only the group of forms which used to be included in the section Euavena when the genus was considered to include certain grass species. The genus is consequently different in structure from *Hordeum*, and certainly also from *Triticum*, if the most recent interpretation of that genus is accepted, and this is emphasized by the fact that no intergeneric hybrids are known.

Phylogenetically the three characteristic features of the genus are its isolation, the association of true wild forms, weed forms and cultivated forms, and a polyploid series similar to that in wheat—diploid (14), tetraploid (28) and hexaploid (42) (Fig. 23). There is, however, no general agreement on the structure of the genus in terms of the number of species, and although seven species seem to be basic, as many as fourteen are sometimes named (Malzew, 1930; Hector, 1936). The evolutionary sequence depends on the transition from the wild characters to the accepted cultivated criteria based on spikelet characters, and the development of polyploidy. The essential difference between the non-cultivated forms—both wild and weed—and the cultivated forms, is in the shedding characters of the former, combined with adaptation of the shed portion for burying in the ground, and the development of awns, and the grain retention, or non-shedding characters of the latter. Shedding is brought about by articulation of the whole spikelet, or separate articulation or disjunction of the separate grain, while this primitive character is associated to a greater or lesser degree with the development of a pronounced callus and hairiness. Although the wild character complex is more complicated morphologically in oats than in wheat and barley, there is the same basic concept and the same type of transition (Plate XI).

If, for the sake of clarity, it is accepted that *Avena* has fourteen distinguishable species, the distribution of the wild, weedy and cultivated forms on the basis of chromosome number is significant. At diploid level there are four species—*clauda, longiglumis, pilosa* and *ventricosa*—which occur almost exclusively in the wild in the Mediterranean region and seem to have originated there. The two former species have all florets articulating, and the latter two are disseminated by spikelets. On the other hand, *clauda* and *pilosa* have the more primitive type of elongated callus. There is, therefore, parallel variation in these two groups, named the Inaequaliglumis and Stipitatae, but there is very little other genetic variation. The Inaequaliglumis group, *clauda* and *pilosa*, is regarded as the more primitive, but there is obviously very close relationship with Stipitatae forms, *longiglumis* and *ventricosa*.

But the diploids have differentiated beyond this primitive

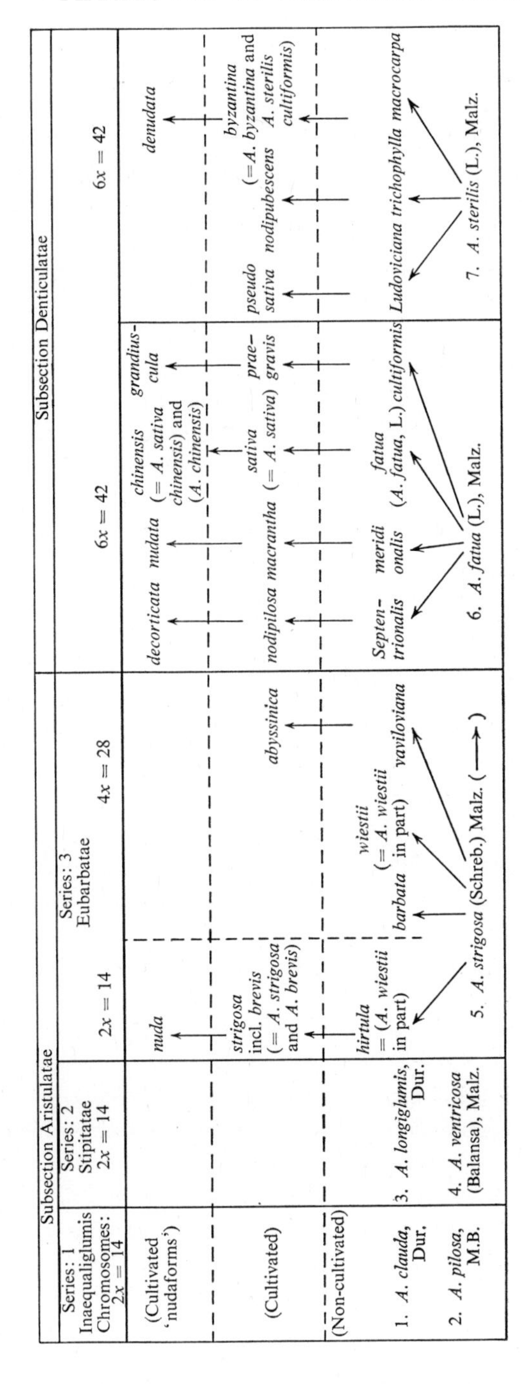

Fig. 23. Cytological relationships of the species of Euavena (after Malzew, modified).

state, as there is a weed species—*hirtula*—with all florets articulating, and also the cultivated species, *strigosa*. *Hirtula* is sometimes regarded as the transitional form to the derived diploid *strigosa*, and, like the other species, the centre of origin would appear to be the Mediterranean. The diploid cultivated *strigosa* has differentiated sufficiently to develop a distinct form (*brevis*), and a naked form (*nuda*), both of which have been cultivated as far north as this country, but have been grown primarily as forage, and are the most ancient and primitive of cultivated oats. The diploids are obviously closely related, they all have the genomic structure *AA* with a slightly modified A^1A^1 in *longiglumis*, which, however, is often said to have closer affinities with *hirtula* than with the other species, and there appears to be an obvious transition from the genetically dominant wild complex of these diploids to the cultivated types.

Diploidy, however, has had obvious limitations in the evolution of cultivated oats, and tetraploidy has determined the next stage in the development of the genus. At this chromosome level, the section Eubarbatae of the genus, the domination of the true wild type is left behind, and the tetraploids may be considered to comprise the wild species *wiestii*, the weed species *barbata* and *vaviloviana*, and one cultivated species *abyssinica*. These *abyssinica* forms are often regarded as the tetraploid level of the species *strigosa*, which includes the diploid *hirtula* and *strigosas* referred to previously. The three non-cultivated tetraploids, apart from *vaviloviana* which occurs only in Abyssinia and Eritrea, have achieved a wider ecological distribution than the non-cultivated diploids, particularly *barbata* which has become distributed throughout the Old World and North America. But the cultivated *abyssinica* is limited ecologically, and has its most suitable environment in Abyssinia, Eritrea and Egypt.

The cytological information indicates that all tetraploids have the basic *AA* chromosome set characteristic of the diploids to which has been added a second set—*BB*. According to Rajhathy and Morrison (1960) the *B* genome has not contributed any chromosome with a distinctive structure, and the origin and status of this genome are not known. Hybridization has shown the complete homology of the *A* of the diploids with the *A* of the tetraploids, and Griffiths *et al.* (1959) demonstrated a close affinity cytologically between the non-cultivated *barbata* and the cultivated *abyssinica*. All the evidence including karyotype analysis would indicate an allopolyploid origin and structure for the tetraploids, and it would seem reasonable to suppose that the one cultivated group arose from *barbata*, though *vaviloviana* has also been quoted as the key non-cultivated form. This species has certain characters, such as the shape of the callus and the method of grain articulation which associate it with the hexaploids rather than the tetraploids. The so-called 'wild complex' is, as in the diploids, dominant to the cultivated complex, and it is assumed that evolution has taken place through structural chromosome changes and mutation. An acceptable picture is offered here of the diploid *AA* set being contributed by a *hirtula*-like form which has been shown to have affinities with *barbata*, and also with the wild tetraploid *wiestii*. On the other hand, although there is obvious evidence for a close relationship between the diploid and tetraploid groups, the structure of the two groups is such that hybrids between them are difficult to obtain.

Nothing is known of the origin of the *B* genome of the tetraploids, but the position is even more obscure for the hexaploids. This group consists of what are usually considered as two basic species each of which has weed forms, cultivated hulled forms, and cultivated naked forms. On the other hand, the weed and cultivated forms of each species are sometimes referred to separate species, thus giving four species. The distinction into the two basic types is made on the somewhat flimsy distinction of the articulation or fracture of the upper grain of the spikelet, although there are other characters including the geographical distribution which can be used. The two cultivated species are *sativa* and *byzantina*, and their weed (or so-called wild) counterparts are *fatua* and *sterilis* respectively. There is much controversy as to the relationship phylogenetically between these forms and a great deal of work has been done on the complexes associated with the grouping (Hector, 1936).

It is probably best to consider first what is known of the genomic structure of the hexaploids in order to visualize at the outset their relationship to one another, and with the diploid and tetraploid forms. The hexaploids possess the basic *A* set of the diploids and tetraploids, but this is all they have in common with the tetraploids. The other two genomes—the *C* and the *D*—are distinct, and their origin is unknown. As Rajhathy and Morrison (1960) have pointed out, this hexaploidy may have been built up in one step—*A* and *CD*—or in two steps—*A* and *C* and *D*—and there is no evidence in support of either view. The important genomic characteristics of oats are that there is no intergenomic homology between the *A*, *B*, *C* and *D* sets; that there is a high degree of homology between the hexaploids; and that there is low homology between the hexaploids and the tetraploids. Consequently, other than that they are allopolyploids, and have a common set of chromosomes with the diploids and tetraploids, the origin and evolution of hexaploid oats is an unsolved problem.

The hexaploids can be considered in terms of what is known as the *sterilis* and *fatua* complexes, in which occur a parallel range of variation from the so-called wild (in effect, weed) type to the cultivated. It has long been considered that the two series evolved separately, as might be suggested by their geographical distribution, the wild *fatuas* having given rise to the cultivated *sativas*, and a type similar to *ludoviciana* having given rise to the cultivated *sterilis* forms, particularly the *byzantinas*, while in each the naked forms are regarded as the ultimate form of differentiation. Both complexes express themselves markedly in terms of well-defined ecological races, but the morphological differentiation is limited, and confined primarily to comparatively unimportant economic characters in terms of grain characters and panicle morphology.

Jones (1956) pointed out a number of years ago, when considering the relationship between the wild and the cultivated forms in the two series, that the genes controlling the wild grain base type are probably homologous, but that the normal grain base type of the hexaploids is distinct and specific to these forms, the effect being to suppress the wild type. In addition the gene concerned is linked with the gene inhibiting awn development. But the interesting and significant feature of the hexaploids is that the grain base character of the cultivated type is genetically dominant to that of the wild *fatua* type, while that of the *sterilis* type is dominant to both. In the diploids and tetraploids, the wild complex is dominant throughout. The position in the hexaploids led Coffman (1946) to propose that the wild *sterilis* is the primitive hexaploid, which gave rise first to the cultivated *byzantina* and then to the cultivated *sativa* groups by progressive solidifying of the basal grain articulation and by definition of the point of fracture of the upper grain. This theory would mean that the *fatua* wild oats are derived forms and not primitive. It also gives due importance to the occurrence in the *sterilis* cultivated forms of a wide range of dominant characters with the occurrence of extreme physiological types and much disease resistance.

It does not seem that there is sufficient evidence to come to any definite conclusion regarding the ultimate differentiation of the hexaploid forms in terms of true phylogenetic relationships. The concept of a basic *sterilis* type as the progenitor of the wide range of non-cultivated and cultivated *sterilis* forms, and also of the equally wide range of non-cultivated and cultivated *fatua* forms, obscures certain important aspects of what would appear to be a parallel differentiation and evolution in the two groups. This would suggest, when one considers also the geographical distribution of the two groups, an early separation of the two systems. On the other hand, one must presuppose an initial common origin if chromosome homology means anything.

The four non-cultivated sub-species of *fatua*, which have been recognized by Malzew represent four distinct ecotypes with an extensive geographical range, and these are associated with cultivated sub-species covering a vast area stretching across the northern hemisphere. On the other hand the three recognized non-cultivated sub-species of *sterilis* are distributed in Asia Minor, the Mediterranean, and the Mediterranean through Afghanistan to South-west Asia. These are associated with the Asia Minor, Mediterranean and Switzerland cultivated forms respectively. The ultimate differentiation in oats is predominantly physiological and ecological rather than morphological, and probably stems from an original South-west Asia centre for *fatua* and possibly also for *sterilis* but with a subsequent different ecological bias. It is interesting to note that Helbaek concludes that oats as a cultivated crop is European, never having been cultivated in the Near East, the first signs of cultivation being Switzerland, Germany and Denmark at the beginning of the first millennium B.C.

The exploitation of the genetic variability of oats has taken place very largely at hexaploid level, there being little to attract attention in the tetraploids, and still less in the diploids. Nevertheless, breeding work has been conducted on a very restricted scale at diploid level for the improvement of forage oats for special conditions, but they cannot compete as grain producers with the tetraploids, and still less with the hexaploids, under good farming conditions. The tetraploid *strigosas* (*abyssinica*) are also of restricted use, but they have been used in hybrids with hexaploids in some breeding work, but with limited possibilities. The hexaploids, as in the wheats, provide the most promising opportunity for oat improvement, and work has been concentrated within the *sterilis* and *fatua* groups, while great use has been made of hybrids between the two. The result is that the two divergent streams of hexaploid oat have been brought together in the improvement of the crop by hybridization between the cultivated forms, and, in particular, use has been made of the more extreme expressions of ecological adaptation in the *sterilis* group for the improvement of the *sativas* or cultivated *fatuas*. In addition, there has been recourse to the resistance to powdery mildew found in the wild *sterilis* form *ludoviciana* which is a widespread weed of cultivated land, thus showing an interesting parallel with the use of *spontaneum* for the same purpose in barley.

Rye

Rye offers several unique features, and is the nearest relative of wheat among the cultivated cereals, the genus (*Secale*) being considered to be very close to *Triticum* taxonomically and phylogenetically. The cytological evidence, however, does not indicate that the rye genome is closely related to that of any other genus in the Triticeae, although hybrids have been made with *Triticum*, *Agropyron* and *Aegilops* in the Triticinae, and with *Hordeum* and *Haynaldia* in the Hordeinae. There is, however, no suggestion of any recent association even of the *Triticum* and *Secale* complexes.

Plate V. Diploid wheat. Right, the wild or weed species *Triticum aegilopoides*. Left, the cultivated species *T. monococcum* ('Einkorn').

facing p. 96

Plate VI. Tetraploid wheat. Right, the wild or weed species *Triticum dicoccoides*. Left, the cultivated species *T. dicoccum* ('Emmer').

Plate VII. Synthesis of hexaploid wheat by allopolyploidy. Reading from left to right—*Aegilops squarrosa*; *Triticum dicoccoides*; synthetic alloheaxploid resembling *T. spelta*; a form of *T. spelta*.

Plate VIII. Hexaploid wheat species. Reading from left to right—*T. spelta; T. vavilovii*; *T. macha*; *T. sphaerococcum*; *T. compactum*; *T. aestivum*.

Plate IX. Wild or weed barley. Left, *Hordeum agriocrithon* (six-row type); right, *H. spontaneum* (two-row type).

Plate X. Major variation of ear type in cultivated barley, showing parallel variation in two-row forms (upper row) and six-row forms (lower row).

Plate XI. Wild, weed and cultivated oats, showing differentiation at diploid, tetraploid and hexaploid levels. Reading from left to right—
Top row (diploids): *Avena pilosa; A. longiglumis; A. hirtula; A. strigosa.*
Middle row (tetraploids): *A. wiestii; A. barbata; A. abyssinica.*
Bottom row (hexaploids): *A. ludoviciana; A. byzantina; A. fatua; A. sativa.*

Plate XII. Rye. Reading from left to right—*Secale cereale*; wild species typified by *S. montanum*.

The genus is small and restricted in its genetic variation, existing as wild grass species, weed species and a cultivated species (Plate XII). This is comparable to *Hordeum*, and in *Secale* it is an acceptable practice to divide the genus into an Agrestis section and a Cerealia section. Morphologically the main characteristic is the extreme uniformity of the ear characters, the occurrence of perennial forms in the grasses, and the transition from the wild, brittle rachis forms in the grasses to the tough rachis forms in the Cerealia, accompanied also by a reduction in the number of flowers in the spikelet.

The genus is recognized as variously having from five to fourteen species, and if accessory chromosomes are disregarded, there is only the single basic chromosome number of $2n = 14$. The most recent review of the genus by Jain (1960) reduces the genus to five species on the basis of cytological evidence and interspecific hybridization. There is little doubt there are two distinct groups of species, and Jain concludes that *Secale* is an expanding genus with speciation having so far taken place by structural chromosome changes and genetic differentiation.

The work on interspecific hybridization in *Secale* has really not progressed far enough to understand the phylogenetic relationships between the species. It is, however, fairly clear that the cultivated *cereale* chromosome complement is differentiated from that of the *montanum–africanum* complement by two reciprocal translocations, and this difference is responsible for a sterility barrier which, according to Price, is at an incipient stage of development. It has been suggested that morphological characters associated with individual species differentiation are being preserved by these translocations, but this does not appear to be certainly true in this material. Interspecies crosses have, however, demonstrated the dominance of certain primitive characters such as rachis fragility, hairy neck, and perennial habit which varies in its dominance expression. The behaviour of the self-compatible and self-incompatible character in interspecific crosses unfortunately has not been thoroughly worked out, so that the relationship between the compatibility of the *africanum–sylvestre* group, and the incompatibility of the *cereale–montanum* group is not clearly understood.

It is possible to generalize on the position in rye by concluding that there is more variability in the wild forms than in the cultivated, both morphologically and physiologically, and once more there is the loss of the primitive wild and weed characters—fragility of the rachis, investment of the grain, etc.—but there is a large number of intermediate forms. The wide area of occurrence of the wild and weed forms does not coincide with the major areas of cultivation of the crop. The wild characters have been selected out and the crop has followed the migration of cultivated wheat and barley, and in spite of its homogeneity, cultivated rye does show a separation into distinct geographical groups, or races. The out-pollinating nature of the crop has resulted in the quantitatively differentiated populations, and there is a unique lack of distinction into discrete botanical forms and agricultural varieties.

Rye, like oats, is a secondary crop with a history in cultivation which probably dates back only as far as the first millennium B.C. Wild rye is native to Turkey and Afghanistan, but cultivated rye has achieved its importance primarily in the cold temperate regions of cereal cultivation, its most valuable characteristics being its extreme winter-hardiness, resistance to drought and capacity to grow on light, sandy and acid soils. In this respect, rye is often regarded as a poverty crop, and although it has served, and continues to serve, a most valuable function in agriculture, it is interesting that whenever it is practicable, there is a swing back to wheat cultivation. This, of course, is due to the preference for wheat as a bread corn and the trend is emphasizing in a most interesting way what may turn out to be the most valuable use of rye in the improvement of the temperate cereals—its possible exploitation in wheat breeding.

Wheat-rye hybrids have been studied for many years, and the so-called *Triticale* amphiploids have received a great deal of attention, particularly the fifty-six chromosome forms, although some attention has been given to those with forty-two chromosomes. However, there does not appear to be true homology between any wheat chromosome and any rye chromosome, and the chance of developing, by accepted conventional breeding methods, wheats with specific rye characters is very remote as pointed out by Riley and Chapman (1958*a*). But there is another possibility of exploiting in wheat breeding the phylogenetic relationship of wheat and rye, and this is by the transference, by addition or substitution, of individual rye chromosomes to wheat. The phenotypic changes resulting from the introduction of individual rye chromosomes are largely quantitative, showing their effects in the size or rate of development of the plants' organs. Qualitative effects in terms of resistance to powdery mildew and yellow rust and ear morphology are, however, also apparent, and there is every reason to continue attempts to exploit this technique.

Conclusion

It has not been possible to consider Vavilov's classical work on centres of origin and centres of diversity, and his concept of the law of homologous series in variation, and on parallel development and evolution as applied to species, genera and even families. But Vavilov did make a most significant statement, among many others, to the effect that variability is primitive in cultivated plants, and this has surely been borne out by work in other crops, as well as what we can see in the Gramineae. Only a very small number of plant families has contributed a wide range of cultivated forms, and a restricted number of genera dominate the scene as far as agriculture is concerned. Of these families, the Gramineae is outstanding, and of these genera *Triticum*, *Hordeum*, *Avena* and *Secale* compose the entire picture as typical temperate cereals. The Gramineae is a dominant family in agriculture, and of course it is regarded as what is known as a successful family. In other words, man has been able to exploit the natural variability of this successful family, and direct its prodigality to his own uses.

The nature and distribution of natural variability of cultivated plants is, consequently, of vital interest to man and in particular to those who are trying to improve cultivated plants. This, of course, is closely bound up with evolution and phylogenetic changes and relationships, and it may be said that plant breeding methods and techniques are simply modelled on evolutionary processes, and the manner in which genetic variation has arisen and new species have developed. In this respect, Anderson's ideas (1953) on character association analysis as a useful tool for the breeder are valid, and his hypothesis that the natural variability of wild populations is almost exclusively the result of introgression—that is, of occasional hybridization followed by extensive backcrossing—is most significant.

The significance of hybridization in the evolution of cultivated plants is generally accepted with regard to its rôle in providing the bases for the release of new bursts of genetic variability, and also because of possible mutagenic effects resulting from certain types of distant hybridization (Harlan, 1961). Introgressive hybridization involving distinct species, or wild, weed and cultivated forms, has obviously been a vital factor in the evolution of cultivated cereals. From what can be seen, the position of the weed forms in the evolutionary processes involving the differentiation of the cultivated cereals, is of the utmost significance, and, indeed, Harlan ascribes an immensely important rôle to the weed forms. It seems clear that cycles of introgressive hybridization must be involved with the weed forms developing greater diversity and acting as sources of new variation, as is seen in the Israeli barleys today. Differentiation by macro-mutations is, of course, inherent in the evolutionary process, and has been a potent force as much in polyploid diversification as in diploid. This would appear to be true of the tetraploid and hexaploid wheats, as well as of barley with its diploid structure.

Now, as Anderson has pointed out, complexes of associated characters persist after introgression, even when there is strong evidence that such introgression occurred as long ago as Tertiary or early Pleistocene. But in terms of man's history it seems probable that the tremendous range of cultivated crops, and the more restricted range of the secondary crops, have developed in a comparatively short time. Harlan, supported by Helbaek's evidence, considers for example that einkorn, emmer and barley were intermediate between the wild and domestic conditions at the beginning of the fifth millennium B.C. At all events, it would seem probable that all the cultivated cereals were domesticated during the agricultural developments of the Neolithic age. It might also be assumed that during the late Stone Age the primary cereals were quite well advanced in their domesticated history, but oats and rye were still weeds.

The position as we know it today from the improved knowledge of man's history, of the phytogeography of the cultivated plants, and the exciting developments in genetics and cytology, is one both full of promise and humbling for the plant breeder. Everyone interested in the improvement of our major cultivated plants is concerned about the maintenance of the basic genetic variability of these plants, and it is significant that this variability is the result of spontaneous evolutionary forces resulting in the great multiplicity of forms which have been maintained in cultivation by the most elementary processes of selection. The major problem is, however, not to collect and maintain the infinite number of phenotypes, but to ensure that potentially valuable genes are not lost. It is as possible parental material that the value of the genetic variability shows itself, and in the crops we have been considering, it is in hybridization programmes that the exploitation can be practised. The problem in planning breeding programmes, centred on the improvement of individual crops for specific purposes, is to ensure that in the first instance they are based on the best genetic material, and that the methods employed are giving the breeder the best chance of achieving the desired results.

It must be accepted that in these cereal crops evolutionary processes have moulded the range of genetic variability to the point where there is not likely to be any fundamental or revolutionary change in the species concerned. Botanical studies and genetical investigations have indicated the range of characters available and those that can be used for the improvement of the crops concerned. Although we still do not know the full genetical diversity of all these cereals, the important investigations that have been going on in recent years are cataloguing the genes and indicating the best sources for use in breeding programmes in various parts of the world. New genotypes are being constituted by breeders in considerable numbers, but these genotypes are mostly derived from a comparatively narrow range of cultivated varieties within closely related groups. There are examples of the same breeding material being used in many parts of the world, while it is common practice for improved and highly developed new varieties to be used as the basis of new breeding programmes in several countries. There is inevitably, therefore, a tendency to narrow the genetic variability being used, with the possible loss of valuable characters.

This procedure may go on indefinitely within the framework of accessible breeding material and great advances are being made thereby in individual crops in various parts of the world. One is inevitably faced, however, with the speculation of how long this type of breeding work will match the requirements of agriculture which will have to meet a greater pressure as world food problems become more and more acute. The four cereals offer different prospects for bringing about major and radical changes, but only in wheat can one really visualize developments of the kind that took place under the circumstances of active evolutionary change which were responsible for the main steps in the synthesis of the crop.

It is problematical whether these cereals have evolved too far for man to be able to retrace certain critical steps in an endeavour to recreate on different lines a spectrum of cultivated forms that will meet more satisfactorily the requirements of the future. Breeding and genetic work so far indicate that the limitations of this sort of approach are very definite with the methods available. Nevertheless, methods and techniques improve, knowledge expands, and new horizons show themselves. We can at least claim that conscious improvement by directed methods has made a great contribution, and no doubt there is every reason to suppose this can continue. The question is, will this be sufficient?

V

CYTOGENETICS AND THE EVOLUTION OF WHEAT

by RALPH RILEY

THE study of comparative morphology and physiology, of ecology and plant geography, of archaeology and philology, can effectively contribute to our understanding of the evolutionary development of crop plants. However cytogenetic investigations must invariably provide the most rewarding sources of exact information. For, built into the contemporary genetic and cytological structure of our crops, there are traces of earlier systems of cytogenetic organization. These await the application of continually improving techniques, and advancing sophistication, by succeeding generations of investigators, to be resolved with increased clarity. With each access of new evidence a better approximation to a true picture of the evolution of crops is attained.

The story of the investigation of the common wheat of agriculture, *Triticum aestivum*, provides an excellent example of the application of cytogenetics in evolutionary studies. *T. aestivum*, which is an inbreeding species, is better understood cytogenetically than any other polyploid organism, and some of the methods applied and results achieved in the exploration of its cytogenetic structure can also be used to illuminate its evolutionary history. The purpose of this contribution is to discuss the evidence for the three major steps—two advances in polyploidy and the achievement of a balanced polyploid organization—which have led to the present pre-eminence of wheat.

No attempt will be made to assess the many adjustments of small individual effect which must also have been involved. While the significance of such modifications should not be ignored they can be regarded as having been subsidiary to, and dependent upon, the major steps. To begin with, therefore, attention will be concentrated on the processes of polyploid advance.

Information on the broad outline of the evolutionary history of *Triticum aestivum* has resulted from the application of the processes of genome analysis. Indeed it was primarily in the wheat group that genome analytical methods were developed by Kihara and his collaborators (Lilienfeld, 1951).

In the sense in which the term is used here a genome may be defined as the haploid set of chromosomes of a diploid species or group of species. Species which have diverged to such an extent, in chromosome structure and gene content, that meiotic conjugation is considerably impaired in hybrids between them, may be said to possess different genomes. It is conventional to attach distinctive symbols to the genomes recognized in this way.

Genome analysis involves the determination of the parentage of an allopolyploid species by comparisons of the extent of meiotic chromosome pairing in hybrids between the polyploid and a range of genomically distinct diploid analysers. However, whilst the system of recognition and notation provides a ready method of describing evolutionary and experimental situations, it is important to be aware that the genic and structural differences between the chromosome sets distinguished may vary widely.

As was first established by Sakamura (1918), the genus *Triticum* consists of a polyploid series of species in which there are diploid, tetraploid and hexaploid representatives with chromosome numbers of 14, 28 and 42 respectively. The basic number of the genus, and the number of chromosomes in each genome, is thus seven. At each level of polyploidy there are taxonomically distinct forms which have been assigned specific or subspecific rank according to the predilections of the taxonomist concerned (Schiemann, 1948; MacKey, 1954*b*; Bowden, 1959). All the forms at each level of polyploidy hybridize readily and the resulting hybrids have normal chromosome conjugation and are fully fertile. Consequently all the forms with the same chromosome number may be regarded as having similar genome constitutions. The economically important bread wheat, *T. aestivum*, may properly be regarded as a genomically typical representative of the hexaploid group. The only exception to this pattern of behaviour is the tetraploid *T. timopheevi* which forms sterile, meiotically irregular, hybrids with the other tetraploids. However *T. timopheevi* has recently been shown to match the other tetraploids more closely than had been supposed hitherto (Wagenaar, 1961).

The relationships between the diploid, tetraploid and hexaploid species of *Triticum* can be determined from the meiotic behaviour of hybrids (Kihara, 1924). Before discussing such hybrids, however, it should be mentioned that the haploid individuals, which occasionally arise parthenogenetically at all levels of polyploidy, have very little chromosome pairing at meiosis. Consequently any considerable conjugation observed in hybrids must result from the association together of chromosomes from different parents.

As a reference point from which to describe a series of comparisons it is convenient to call the haploid set of seven chromosomes of the diploid wheats the *A* genome. Hybrids between diploid and tetraploid forms usually have seven bivalents and seven univalents at meiosis, showing that the tetraploids have the *A* genome of the diploids plus a further genome—the *B* genome. Hybrids between tetraploids and hexaploids have fourteen bivalents and seven univalents at meiosis, so that the hexaploids have the *AB* genomes of the tetraploids plus one further genome—the *D* genome.* Hybrids between diploids and hexaploids commonly have seven bivalents and fourteen univalents, demonstrating the general similarity of their *A* genomes, but the lack of exact equivalence is indicated by the occurrence of cells with less than seven bivalents.

From these hybrid comparisons the relationships within the genus can be shown as follows:

$$\begin{aligned} &Triticum\ 2x = 14 = AA \\ &Triticum\ 4x = 28 = AABB \\ &Triticum\ 6x = 42 = AABBDD \end{aligned}$$

The evolutionary implications of this situation clearly are that the tetraploids arose as a result of hybridization between diploid wheat and the *B* genome diploid species. This hybridization was accompanied or followed by doubling of the chromosome

* It is an accident of nomenclature of the related diploid species that the third wheat genome has received the letter *D*, and not *C*.

number. Subsequently hybridization took place between tetraploid wheat and a third diploid species, the *D* genome donor. Doubling the chromosome number in this hybrid gave rise to the hexaploids and so, either directly or indirectly, to *T. aestivum*.

The diploid forms of *Triticum* represent a fairly compact genetic group in that there is no suggestion of internal divergence of sufficient magnitude to impair hybrid fertility. Yet it might be possible to specify from which sector of the range of diploid variation the *A* genome of the polyploid wheats was extracted. Riley and Bell (1958) have indicated certain genetic and cytological evidence which seems to imply that *T. thaoudar*, the most primitive of the diploids, is the contemporary form closest to the *A* genome donor. It should be stressed that this is an extremely tentative conclusion, but if true it would mean that the first tetraploid *Triticum* probably did not arise in cultivation, since both the parents and their derivative are wild-growing species at the present time.

The source of the *D* genome, which is found in the hexaploid but not in the tetraploid wheats, was discovered in the first instance from work with the allotetraploid species *Aegilops cylindrica* ($2n = 4x = 28$). Sax and Sax (1924) demonstrated that hybrids between *A. cylindrica* and hexaploid forms of *Triticum* usually have seven bivalents and twenty-one univalents at meiosis. Subsequently Bleier (1928) showed that hybrids between tetraploid wheats and *A. cylindrica* rarely have any chromosome conjugation at meiosis. Thus it was clear that *A. cylindrica* carried the *D* genome of hexaploid *Triticum*.

Genome analysis of *A. cylindrica* showed that it carried the *C* genome of *Aegilops caudata* ($2n = 14$) and the *D* genome of *Aegilops squarrosa* ($2n = 14$) (Kihara, 1937; McFadden and Sears, 1946). Since there was no chromosome pairing in the hybrid between *A. caudata* and hexaploid wheat, it could be concluded that the *D* genome of *A. cylindrica* was present at the hexaploid level in *Triticum*. Therefore *A. squarrosa* must have contributed the *D* genome to *T. aestivum* and the related hexaploids.

The climax of the *D* genome story was reached when McFadden and Sears (1946) succeeded in making the hybrid between tetraploid wheat and *A. squarrosa*. The synthetic hexaploid, which resulted from doubling the chromosome number in this hybrid, closely resembled the hexaploid form *T. spelta* in gross morphology. Moreover there was regular bivalent formation at meiosis in hybrids between the synthetic hexaploid and natural hexaploid wheats, and the hybrids were fertile. Thus it could be regarded as well established that *A. squarrosa* had supplied the *D* genome.

The final stage in the investigation of this genome was reached when, for the first time, a hybrid was obtained between *A. squarrosa* and *T. aestivum* (Riley and Chapman, 1960*b*). In view of the relationships between the species concerned, this hybrid is peculiarly difficult to make, but it was ultimately produced by culturing a hybrid embryo on an artificial medium. The meiotic behaviour of the hybrid was a striking confirmation of the presence, in hexaploid wheat, of the genome of *A. squarrosa* (Table 2, Plate XIII), since in the majority of cells there were seven bivalents and fourteen univalents.

TABLE 2. *Chromosome pairing at first metaphase of meiosis in the hybrid* T. aestivum *var. Chinese Spring* × A. squarrosa

	Pairing	
Number of cells	Biv.	Univ.
60	7	14
35	6	16
4	5	18
1	4	20
Mean (100 cells)	6·54	14·46

As a result of the accumulated evidence, discussed above, there can be no doubt that the hexaploid forms of *Triticum* arose from a hybrid between a tetraploid wheat and *A. squarrosa*, in which the chromosome number was doubled. The present distribution of *A. squarrosa* extends from Afghanistan and northern Pakistan, through central-southern Asia, to the eastern tip of the Black Sea. Throughout this area tetraploid wheat also occurs, and it is perhaps significant that *A. squarrosa* is predominantly a weed of wheat fields (Kihara and Tanaka, 1958). Indeed the Kyoto University Scientific Expedition of 1955, found that in certain samples of wheat, sold for chicken feed in Chalus on the Caspian

Sea, the frequency of spikelets of *A. squarrosa* was as high as one to every four grains of wheat. Consequently it seems that the initial hybridization may well have occurred under conditions of cultivation. If this were so the first hexaploid wheat may have been a free-threshing type resembling *T. aestivum*, as suggested by Kihara and Lilienfeld (1949), rather than a form like *T. spelta* with grains closely invested by the chaff, as proposed by McFadden and Sears (1946). Unfortunately, in the absence of direct evidence, it is only possible to conjecture as to the nature of the initial hexaploid. Similarly the related question concerning the type of tetraploid *Triticum* from which the hexaploids developed cannot be answered from our present knowledge.

The discovery of the origins of the *B* genome of *Triticum*, by genome analysis, would have required the recognition of a diploid species which, when crossed with diploid wheat, gave hybrids with little meiotic pairing but which, when crossed with tetraploid wheat, gave hybrids with considerable pairing. In all the extensive hybridization work carried out between species in the genera, *Triticum*, *Aegilops*, *Secale* and *Agropyron*, no such species has been found. Consequently it became apparent by the 1950's that the source of the *B* genome would probably never be discovered by genome analytical methods (Sears, 1959).

McFadden and Sears (1946) proposed that the characters which distinguished tetraploid from diploid wheats could have been derived from *Agropyron triticeum* ($2n = 14$) (= *Eremopyron triticeum*). However there was overwhelming evidence against this since the chromosomes of all the species of *Triticum* have median or submedian centromeres (Plate XIV), whereas Avdulov (1931) had already shown that the karyotype of *A. triticeum* consisted entirely of chromosomes with subterminal centromeres.

The first major piece of evidence concerning the origin of the *B* genome was produced by Sarkar and Stebbins (1956) using Anderson's (1949) method of extrapolated correlates, on diploid and tetraploid wheat species. They applied the principle, verified from many examples of synthetic allopolyploids in the Triticinae, that polyploids are intermediate between their parents in most characters. It was argued that, from a systematic comparison of the morphology of diploids and tetraploids, it should be possible to predict the characters contributed by the unknown *B* genome parent.

The two diploid forms *T. monococcum* and *T. aegilopoides* were compared with the tetraploids *T. dicoccoides*, *T. dicoccum* and *T. durum* over a range of characters which included:

the shape of the rachis internode segment,
the maximum number of fertile florets per spikelet,
the number of keels per glume,
the number of veins per glume,
the shape of the glume apex,
the texture of the glume margin,
the distribution of lemma awns,
the shape of the lemma apex,
the presence or absence of a split in the palea.

By determining how the tetraploids differed from the diploids in these attributes it was possible to extrapolate the characteristics of the other parent which in combination with diploid *Triticum* gave rise to the tetraploids.

The extrapolate matched, to a remarkable degree, the wild diploid species *Aegilops speltoides* ($2n = 14$) and the hypothesis was confidently advanced that this species was the *B* genome donor. Sarker and Stebbins assumed that the chromosome structure of the polyploid had undergone such modification that *A. speltoides* could no longer be recognized as the parent of polyploid wheat by meiotic pairing in hybrids. *A. speltoides* is in the Sitopsis section of the genus which also includes two other species *A. bicornis* ($2n = 14$) and *A. longissima* ($2n = 14$), as well as a sub-specific form of the latter usually called *A. sharonensis*. It was pointed out by Sears (1956), immediately following the work of Sarkar and Stebbins, that the synthetic tetraploid combining diploid wheat and *A. bicornis* was morphologically closer to tetraploid wheat than was the tetraploid derived from diploid wheat and *A. speltoides*. Thus the immediate choice of the ancestor of the tetraploid seemed to lie between *A. speltoides* and *A. bicornis*.

The problem of the *B* genome was not finally resolved until full information became available on the distribution of all the chromosomes of *T. aestivum* to its three component genomes

(Sears 1958). *T. aestivum*, like all polyploid wheats has two pairs of chromosomes, numbered I (1B) and X (6B), with prominent secondary constrictions which delimit very clear terminal satellites (Plate XIV). Riley, Unrau and Chapman (1958) recognized that Sears' demonstration that both these satellited chromosomes were in the *B* genome provided the possibility of karyotypic identification of the source of this genome. In the event, when the karyotypes of all the likely diploid species were examined, only two were to be found in which there were two chromosome pairs with prominent satellites. These species were *Aegilops mutica* ($2n = 14$), which on the grounds of gross external morphology is unlikely to be the *B* genome donor, and *A. speltoides*. The other species of the Sitopsis section, *A. bicornis* and *A. longissima*, could be eliminated as possible parents since they have only one pair of chromosomes with large satellites, like wheat, and one pair with much smaller satellites (Riley, Unrau and Chapman, 1958). Therefore, from the evidence of extrapolated correlates and of karyotype, it seems highly probable that *A. speltoides* is the parent from which the *B* genome of the tetraploid, and hence of hexaploid, *Triticum* is derived. Moreover, utilizing information on the genetic control of the regulation of meiotic chromosome pairing (Riley and Chapman, 1958*b*), to be discussed later, it is possible to show why the significance of this species could not be recognized by genome analysis (Riley, Unrau and Chapman, 1958).

A. speltoides is distributed in nature in a broad belt from the southern shore of the Black Sea, through central Turkey to the eastern shore of the Mediterranean, along which it extends into southern Israel. The wild as well as the cultivated diploid forms of *Triticum* also occur in this region which must consequently be regarded as the place of origin of tetraploid wheat. In view of the occurrence of a wild tetraploid, *T. dicoccoides*, and of the evidence discussed earlier that the *A* genome of the tetraploids was derived from a wild, rather than a cultivated, diploid *Triticum*, the inference must be strong that the first tetraploid arose outside, even prior to, cultivation. Indeed it may even be that the wild diploids and tetraploids were taken into cultivation simultaneously.

The origins of hexaploid wheat, including *T. aestivum* the

species of supreme economic significance, may therefore be seen to have depended upon two important processes. First there was genetic divergence amongst a range of diploid forms originally derived from a common prototype of much more restricted genetic diversity. This divergence resulted in the development of a number of species each adapted to particular ecological and geographical environments in the region of south-western and central-southern Asia.

Subsequently three of these species, a diploid *Triticum*, *A. speltoides* and *A. squarrosa* contributed to the parentage of hexaploid wheat (Fig. 24, Plate XV). Hybridization between diploid *Triticum* and *A. speltoides* occurred at some stage after the attainment of such divergence in chromosome structure and gene content that the hybrid was sterile because of unbalanced meiotic pairing. An improvement in meiotic balance and the restoration of fertility resulted from a spontaneous doubling of the chromosome number in the hybrid. From this original tetraploid the whole range of current genetic variation at the 28-chromosome level subsequently developed. There are no grounds for postulating that this combined hybridization and chromosome doubling event happened more than once. Indeed the conjunction of these events, which are rare in themselves, with an equally rare mutational event, to be discussed later, emphasizes the extreme probability that all the tetraploid forms of *Triticum* are of monophyletic origin.

A further hybridization between tetraploid wheat and the equally divergent diploid, *A. squarrosa*, again followed by doubling of the chromosome number, gave rise to the hexaploid forms, including *T. aestivum* (Fig. 24). There is no evidence for more than one origin of the hexaploids but the possibility of polyphyletic origins cannot be dismissed with the certainty that it can for the tetraploids.

The phylogeny of *Triticum* may thus be considered to involve the sequential processes of diploid divergence and polyploid convergence. Apparently each step in the polyploid evolution of the genus has permitted adaptation to new environmental situations and has resulted in the extension of its geographical distribution and agricultural utilization.

A striking anomaly in the meiotic behaviour of *T. aestivum* becomes apparent from a consideration of the closeness of the relationships between its three component genomes (Riley and Chapman, 1958*b*). If the diploid species from which these chromosome sets were derived are crossed together in pairs, while there is sufficient failure of meiotic conjugation in the hybrids to permit the recognition of the distinctive *A*, *B* and *D* genomes, nevertheless some pairing does occur. In each of the three hybrids possible between diploid wheat, *A. speltoides* and *A. squarrosa* there is a mean of between three and four bivalents

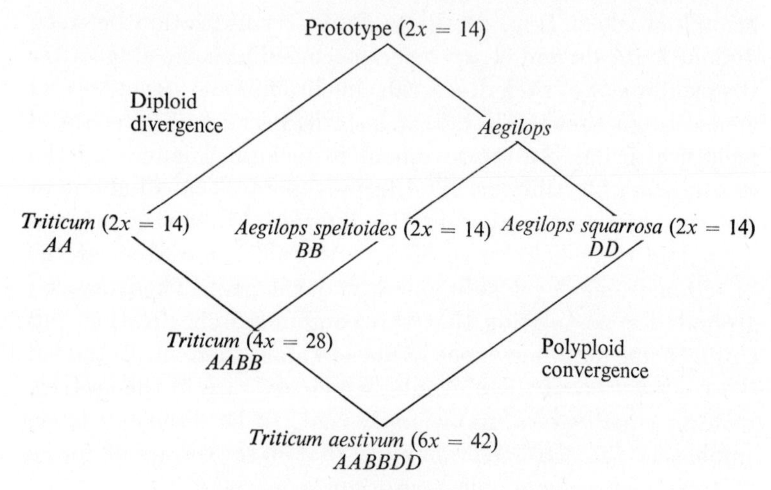

Fig. 24. The phylogeny of *T. aestivum*.

per cell. Moreover there are some cells in which pairing is complete, that is in which there are seven bivalents and no univalents. In addition there is some multivalent formation in the synthetic allotetraploids produced from these hybrids, indicating that even when fully homologous partners are present corresponding chromosomes from the different species are still capable of conjugation. Yet, at meiosis, hexaploid wheat is diploid-like—there are only bivalents and no multivalents. Consequently the capacity of equivalent chromosomes of the three genomes to pair is no longer apparent. Indeed the lack of meiotic affinity is so pronounced that in 21-chromosome haploid individuals, which occasionally arise parthenogenetically, there is little meiotic pairing despite the absence of complete homologues, the presence of which might normally interfere with intergenome pairing. Clearly there has been some modification in meiotic pairing abilities during the evolution of hexaploid wheat.

The anomalous meiotic behaviour was further emphasized by the results of work on the genetic relationships of the chromosomes of *T. aestivum*. Sears (1958) and Okamoto (1962) were able to allocate all the chromosomes of *T. aestivum* to their particular genome by studying the disturbance of meiotic pairing in hybrids between either *AABB* natural tetraploids, or *AADD* synthetic tetraploids, and the 41-chromosome, monosomic, plants deficient in turn for each chromosome of the hexaploid complement. This permitted the classification of the haploid complement into three groups each of seven chromosomes—representing the *A*, *B* and *D* genomes (Table 3).

TABLE 3. *The distribution of the chromosomes of* T. aestivum *to genomes and homoeologous groups* (Sears, 1958; Okamoto, 1962)

Homoeologous group	Genome		
	A	*B*	*D*
1	XIV (1A)	I (1B)	XVII (1D)
2	XIII (2A)	II (2B)	XX (2D)
3	XII (3A)	III (3B)	XVI (3D)
4	IV (4A)	VIII (4B)	XV (4D)
5	IX (5A)	V (5B)	XVIII (5D)
6	VI (6A)	X (6B)	XIX (6D)
7	XI (7A)	VII (7B)	XXI (7D)

The figures and letters in parenthesis are the new symbols proposed for the chromosomes by Sears (1958) following the classification into genomes and homoeologous groups.

In a further classification Sears (1954) used the full set of twenty-one monosomics in combination with the full set of twenty-one, 44-chromosome, tetrasomics. Hybridization between these two kinds of aneuploids made possible the development of plants with reduced dosage of one chromosome but increased dosage of another chromosome. In some instances the abnormalities of the phenotype normally caused by aneuploidy for one chromosome were not apparent when two chromosomes were simultaneously shifted from their normal disomic balance.

The best examples of this behaviour were those in which plants were simultaneously nullisomic and tetrasomic but quite normal in phenotype. The clear inference may be drawn, from such duplication-deficiency compensation, that the chromosome in extra dosage is similar in genetic activity to that for which the plant is deficient.

A remarkable assessment of relationships emerged from Sears' (1958) investigation of a range of material of this type. This showed that the twenty-one pairs of chromosomes of *T. aestivum* could be arranged into seven groups each of three pairs. Within any group there was nullisomic-tetrasomic or some other form of deficiency-duplication compensation, whereas between groups no such compensation was possible. Moreover, when the classification into groups was combined with the classification into genomes, it was found that one chromosome pair in each group was in each genome, and one pair in each genome was in each group (Table 3). This classification, therefore, apparently permits the recognition of the genetically equivalent chromosomes of each genome. The equivalence may reasonably be visualized as tracing back, through the immediate diploid ancestors of *T. aestivum*, to the common prototype of the *Triticum-Aegilops* group. Consequently it is apparently still possible to recognize the genetic correspondence of the chromosomes which are capable of meiotic association at the diploid level. Following Huskins (1931) corresponding chromosomes of the different genomes have been called, 'homoeologous', and there are therefore seven homoeologous groups in *T. aestivum*.

To recapitulate, it may be seen that *T. aestivum* contains the chromosome sets of three diploid species. Moreover the chromosomes of these sets are capable of meiotic conjugation in hybrids at the diploid level. In addition the structures of the chromosomes of the polyploids are not greatly different from those of the parental diploids, since there is no obstacle to the determination of the *A* and *D* genomes through hybrid pairing. Further, homoeologous chromosomes are still alike in some major aspects of their genetic activity. However, despite these indications of genetic and cytological proximity there is normally no meiotic pairing between homoeologous chromosomes in hexaploid, nor indeed in tetraploid, wheat. The difficulties in reconciling these various aspects of the cytogenetic behaviour of *T. aestivum* were resolved by Riley and Chapman (1958*b*) and from this work a new understanding of the evolution of wheat has been achieved.

As has already been indicated, normal 21-chromosome haploids of *T. aestivum*, in which each member of the hexaploid chromosome complement is represented only once, have very little meiotic pairing. There are never more than four bivalents, and the mean frequency is usually between one and two bivalents, per cell. However the study of 20-chromosome nulli-haploid plants (Riley and Chapman, 1958*b*; Riley, 1960) showed that when chromosome V (5B) was deficient there was vastly increased pairing with numerous bivalents and trivalents (Table 4). Apparently chromosome V of hexaploid wheat carries a gene, or a group of genes, which normally limit the extent of meiotic pairing.

It was further shown that in 40-chromosome plants, nullisomic for chromosome V, the usual bivalent-forming meiotic regime was changed to one in which multivalent formation was common. Thus, by the removal of this chromosome the behaviour of *T. aestivum* was converted from that of a typical allopolyploid, with only bivalents at meiosis, to a pattern approaching that of an autopolyploid.

The nature of the pairing in the V-deficient haploids, especially the high frequency of trivalents, was such that it could reasonably be concluded to result from homoeologous conjugation. This inference can be tested by making use of the reciprocal translocations which arise as the result of non-homologous recombination in the nullisomic-V situation. Recent determinations of the chromosomes involved in translocations of this origin demonstrate that they are predominantly between homoeologous chromosomes (Riley and Kempanna, unpublished). There is thus good reason to believe that the original conclusion, that the activity of chromosome V normally prevents homoeologous pairing, is correct.

Before discussing the significance of this device in the evolution of wheat it is necessary to indicate something of the genetic system concerned (Riley, 1960). First, the region of chromosome involved is quite localized. The removal of no other chromosome has an influence on meiotic pairing like that accompanying the removal of chromosome V, so that no other linkage group is concerned. In addition there is no modification of pairing provided that the long arm of the chromosome is present, even though the short arm is absent. By contrast in the reciprocal situation there is homoeologous pairing, so that the activity is restricted to the long arm of one chromosome. In addition there is evidence which indicates a more precise localization within the long arm, so that it is reasonable to think in terms of a single active locus.

The change in meiotic pairing, in the absence of chromosome V, is not due to an alteration in the frequency, or distribution, of chiasmata. That is to say there is no gross increase in pairing, indeed it may be that there is a slight decrease in chiasma-formation. Consequently the normal activity of the effective region of the chromosome must be to modify the processes which initiate synapsis. The system thus depends upon the genetic enhancement of differential affinity.

There is no direct evidence of the evolutionary function of the system but it may be assumed that its importance has been considerable. For by precluding homoeologous conjugation, and confining meiotic pairing to complete homologues, it has made possible the development of that regularity of segregation which is essential to high fertility and genetic stability. It is of considerable importance, therefore, for an understanding of the evolution of the wheat crop, to enquire into the origin of this genetic regulation of the diploid-like meiotic behaviour of the polyploid wheats.

It may be noted (Table 3) that chromosome V is in the *B* genome and was therefore derived either from *A. speltoides* or from some closely related species of the Sitopsis section of *Aegilops*. The first question to be answered is whether the control of meiotic behaviour exercised by this chromosome, in contemporary forms of *T. aestivum*, depends upon some activity which it also performed in the original diploid or whether it results from a change in activity. This was determined by comparisons of the meiosis of hybrids between *A. speltoides*, *A. longissima* and *T. aestivum*, in the presence or absence of chromosome V.

First, however, it should be indicated that *A. speltoides* and *A. longissima*, which are both diploids, are very similar in chromosome structure and hybrids between them have normal meiotic pairing except for the distortion caused by heterozygosity for a reciprocal translocation difference (Riley, Kimber and Chapman, 1961). Plants of *T. aestivum* monosomic for chromosome V were pollinated by the two *Aegilops* species and hybrids obtained which had twenty-seven or twenty-eight chromosomes, depending upon whether or not the irregularly segregating monosomic chromosome was included. These hybrids allowed comparisons of the effects on meiosis of wheat chromosome V and of the genotypes of the two diploids.

In the presence of chromosome V there was little pairing in the hybrids involving *A. longissima*, but in its absence pairing was greatly increased (Table 4, Plate XVI). It could be concluded, therefore, that the genotype of *A. longissima* did not suppress the effects of V, nor could it compensate for the absence of V by preventing homoeologous associations. Thus the *A. longissima* genotype performs no function in the regulation of meiosis which resembles that of the wheat chromosome.

TABLE 4. *Mean meiotic pairing in haploids and hybrids with and without chromosome V of* T. aestivum

Plant	Chrom. no.	Chrom. V	Cells	Univ.	Biv.	Triv.	Quad.	Penta.
T. aestivum haploid	21	present	750	18·84	1·05	0·02	—	—
T. aestivum haploid	20	absent	218	7·42	3·72	1·62	0·07	—
T. aestivum × *A. longissima*	28	present	100	24·08	1·96	—	—	—
T. aestivum × *A. longissima*	27	absent	200	8·63	7·02	0·79	0·49	—
T. aestivum × *A. speltoides*	28	present	100	4·88	6·39	2·04	1·03	0·02
T. aestivum × *A. speltoides*	27	absent	60	6·52	5·67	1·55	1·08	0·03

By contrast there was high pairing in the 28-chromosome hybrid derived from *A. speltoides*. Evidently this genotype suppresses the inhibition of homoeologous pairing exercised by chromosome V. It was not surprising, therefore, to observe no difference in pairing in the 27-chromosome hybrid deficient for V (Table 4, Plate XVI). Since its effect was suppressed in the euploid hybrid, removing it caused no alteration in pairing. Clearly the *A. speltoides* genotype performs a function in the regulation of meiosis completely different from that of wheat chromosome V.

Three quite distinct genetic effects can therefore be resolved:

(a) Chromosome V normally inhibits homoeologous pairing,
(b) *A. speltoides* suppresses the effect of chromosome V,
(c) *A. longissima* does not suppress the effect of V, nor in the absence of V can the *A. longissima* genotype take over its inhibitory function.

It should be added that the other Sitopsis species, *A. bicornis*, behaves in the same way as *A. longissima* when subjected to the same comparisons.

The significant outcome of this work is the conclusion that none of the diploid species, from which the wheat *B* genome might have been derived, produces an influence on meiosis like that caused by chromosome V. Thus the contemporary activity of this chromosome must have resulted from a change in function subsequent to its incorporation in polyploid wheat. The nature of the change has not yet been determined and any notion advanced to explain it depends upon the hypothesis proposed to account for the genetic situation. If the simplest hypothesis is adopted the genetic situation can be explained on the basis of differences at a single locus. The states of this would then be $P_s - p_V - p_l$, where P_s represents the dominant condition in *A. speltoides*, p_V represents a recessive condition associated with chromosome V, and p_l represents the *A. longissima* condition which was of null effect in this test.

If this assessment of the position is correct the chromosome V condition could have arisen by mutation from either the *A. speltoides* or the *A. longissima* state. However, it is *a priori* more likely that the mutation would be from the dominant to the recessive condition, and this coincides with the evidence previously discussed which supports the view that *A. speltoides* was the *B* genome donor. A further confirmation of this is the explanation which can now be offered for the difficulty of determining the *B* genome parent by genome analysis. This results from the confusion caused by homoeologous pairing in hybrids with *A. speltoides*. The improbability of either *A. bicornis* or *A. longissima* carrying the *B* genome can be argued from the general failure of their chromosomes to pair with chromosomes of the *B* genome of polyploid wheat, even though normal genome analysis is possible for the *A* and *D* genomes.

It is then important to enquire at what stage, in the evolution of polyploid wheat, the pairing mutation occurred on chromosome V. Significantly all the polyploid species of *Triticum*, both tetraploid and hexaploid, carry alleles equivalent in effect to that at the pairing locus in *T. aestivum*. The evidence for this is provided by the absence of segregants, with homoeologous pairing, from pentaploid wheat hybrids or from hybrids between different hexaploids. Consequently all the polyploids must have a similar diploidizing mechanism, so that the mutant responsible must have arisen very early in the polyploid history of the genus, presumably at the tetraploid level.

Any discussion of the history of the process of diploidization must be largely speculative but it is reasonable to suppose that the pairing mutant arose soon after the origin of the first tetraploid. The synthetic tetraploid derived from diploid wheat and *A. speltoides*, the probable parents of tetraploid wheat, has multivalents at meiosis and is of low fertility (Riley, Unrau and Chapman, 1958). However it is perfectly stable from generation to generation—displaying none of the segregation which might be expected from intergenome pairing. This must imply that, as the result either of gametic or zygotic inviability, intergenomic recombinants rarely, if ever, survive. This situation has an important bearing on what may be presumed to be the early history of tetraploid wheat for two reasons. First the elimination of the products of intergenomic recombination would permit the maintenance of the individuality of the genomes despite the capacity for intergenome pairing. Secondly there would be a high selective premium on any system which prevented the wastage of gametes or zygotes required by such elimination. It seems likely that the occurrence of a mutation at the pairing factor locus on chromosome V provided the system which allowed tetraploid wheat to escape from this situation. The mutant which gave concomitant increases in genetic stability and fertility presumably had a pronounced selective advantage and quickly advanced to fixation in the tetraploid wheat population.

The same genetic inhibition of homoeologous pairing is now, therefore, found in all the diversity of tetraploid forms which developed from this mutant progenitor. Indeed it is the uniformity of the genetic regulation of meiosis among all the tetraploids which makes it improbable that they are of polyphyletic origin. For this would require the sequential occurrence on more than one occasion of three rare events—hybridization, the development of polyploidy and mutation.

From the tetraploid wheats the chromosome V pairing mutant was presumably passed to the hexaploids, including *T. aestivum*. At this level it would immediately prevent homoeologous pairing between all three genomes. Thus the step from tetraploidy to hexaploidy must have been much simpler than the change from diploidy to stable and fertile tetraploidy. Nevertheless high fertility, the essential requirement of a successful seed crop, has been made possible by the pairing mutant on chromosome V at both the higher levels of polyploidy. It may fairly be claimed therefore that the whole development of the world wheat crop has depended upon this unlikely mutational event.

The three major steps responsible for the evolution of *T. aestivum* have been described. Two of these have involved advances in allopolyploid organization, from diploidy to tetraploidy and then to hexaploidy. Each change in the level of polyploidy has made available a new range of genetic variation which must be regarded as having been primarily responsible for further extensions in geographical and ecological adaptation and in agricultural utilization. For example it seems clear that the incorporation of the *D* genome was responsible for the development of wheats with endosperm suitable for the production of bread-making flours (Yamashita, Tanaka and Koyama, 1957).

The presence of considerable levels of genetic duplication, or triplication, in the polyploids made many loci available at which mutational changes could occur without immediately deleterious consequences. The genetic range open to evolutionary exploration was thus extended and the capacity for precise agricultural adaptation improved. As might be expected from their greater

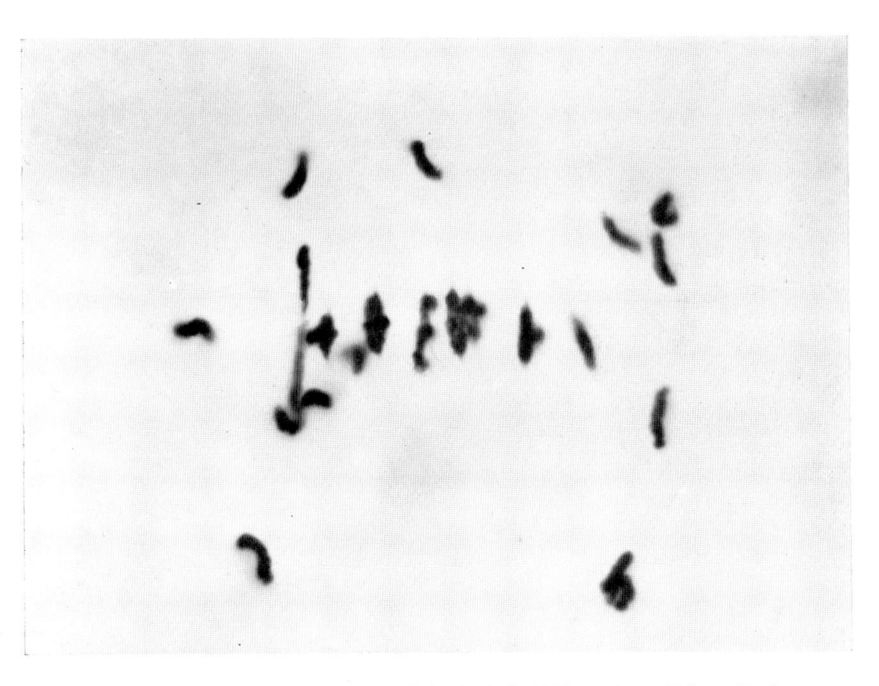

Plate XIII. First metaphase of meiosis in the hybrid *T. aestivum* Chinese Spring × *A. squarrosa*, showing seven bivalents and fourteen univalents. × (1400.)

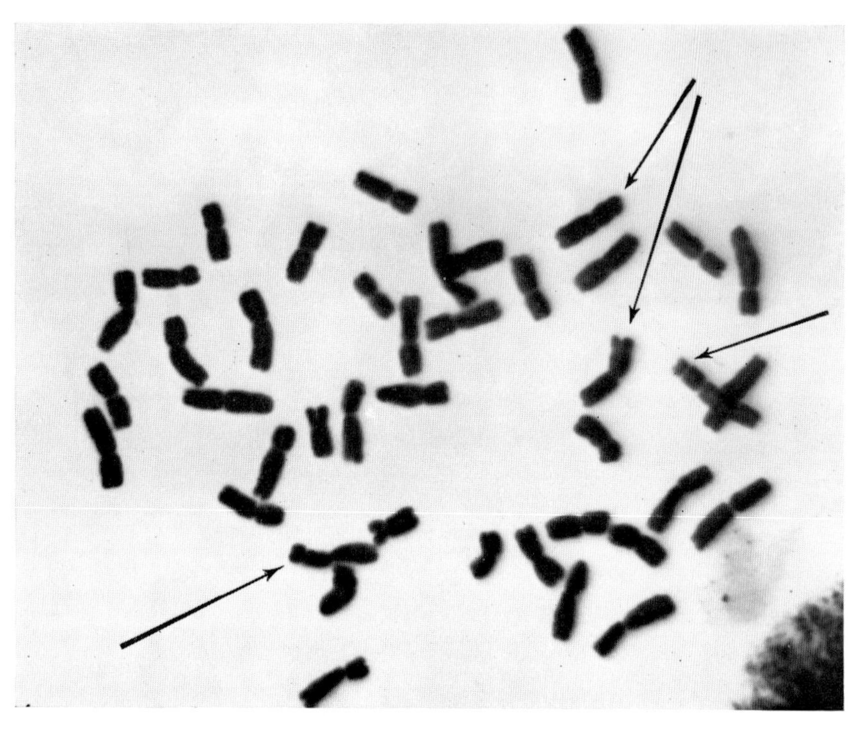

Plate XIV. A root-tip squash, after pretreatment with monobromonaphthalene, to show the somatic chromosome complement of *T. aestivum*. All forty-two chromosomes have median or submedian centromeres and the four satellited chromosomes are indicated by arrows. (×1700.)

Plate XV. Spikes of hexaploid wheat and its three diploid parents, (A) *T. monococcum* ($2n = 14$), (B) *A. speltoides* var. *aucheri* ($2n = 14$), (C) *A. squarrosa* ($2n = 14$), (D) *T. aestivum* var. Chinese Spring. ($2n = 42$).

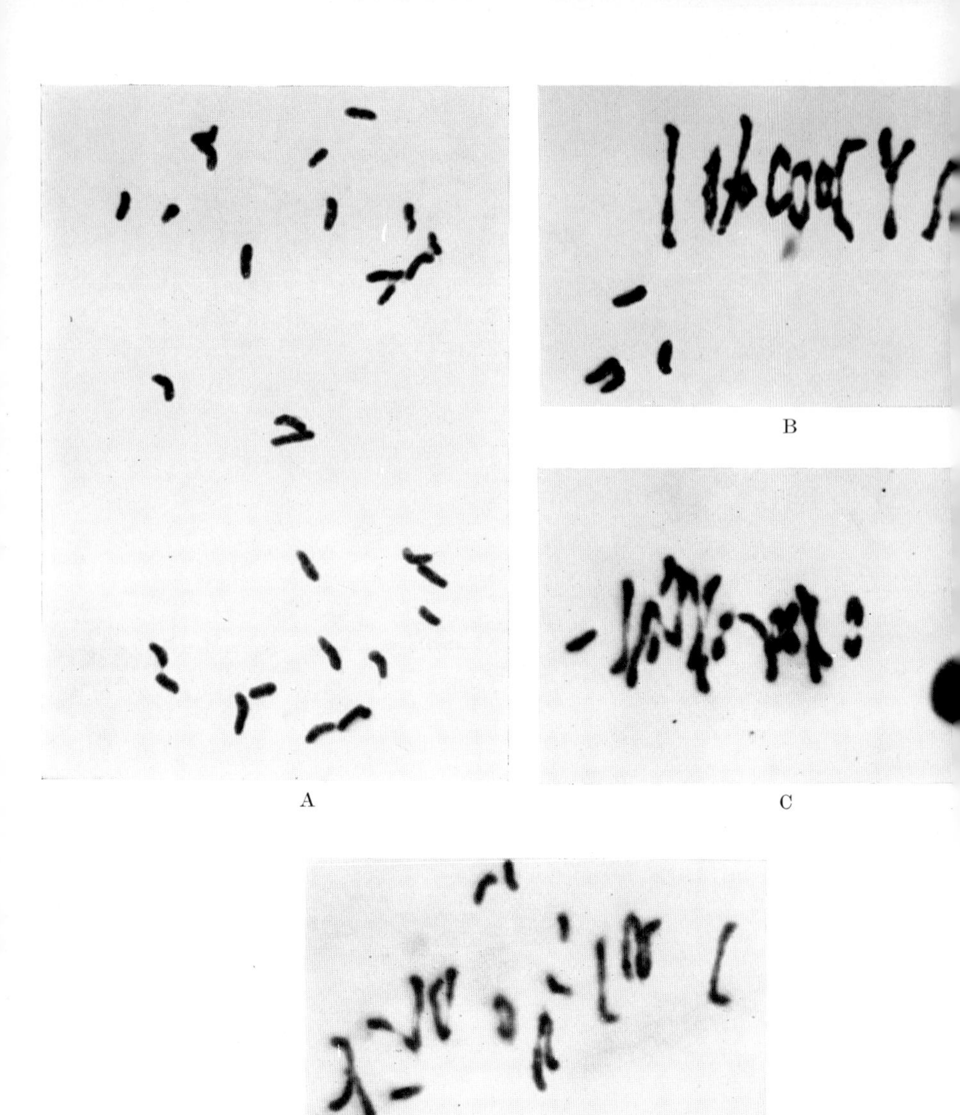

Plate XVI. First metaphase of meiosis in the following hybrids: (A) *T. aestivum* × *A. longissima*, with twenty-eight chromosomes including V of wheat; all the chromosomes as univalents.
(B) *T. aestivum* × *A. longissima*, with twenty-seven chromosomes and deficient for V of wheat; one quadrivalent, one trivalent, 8 bivalents and 4 univalents.
(C) *T. aestivum* × *A. speltoides*, with twenty-eight chromosomes including V of wheat; one quadrivalent, 11 bivalents, 2 univalents.
(D) *T. aestivum* × *A. speltoides*, with twenty-seven chromosomes and deficient for V of wheat; one quadrivalent, four trivalents, three bivalents, five univalents. (All × 1400.)

age, the genetic resilience of the polyploid status has been diminished to a greater extent in the tetraploids than in the hexaploids. By numbers of criteria, which include the tolerance of aneuploidy, the rate of production of vital mutants (MacKey, 1958*b*) and the fertility of pentaploid compared with the sterility of triploid interspecific hybrids, hexaploid wheat has much more genetic duplication than tetraploid wheat.

The diploid-like status of some systems in the genotype of the tetraploids is indicative of the direction of evolutionary change in the adjustment of the polyploid condition. The implication of the presence of much more duplication in *T. aestivum* is that the hexaploids are still in a comparatively unadjusted state. It may be that, in the future movement towards a balance comparable with that of the tetraploids, marked advances can be expected in the genetic exploitation of the crop. Indeed, on the present evidence, an optimistic assessment may be formulated of the genetic potentialities of hexaploid wheat.

However, the adaptive modifications of the genotype which have been possible in polyploid wheat have been dependent upon the third major evolutionary step—the mutation at the pairing locus on chromosome V, which imposed the diploid-like meiotic regime. The disomic system of inheritance, which resulted from the restriction of pairing to complete homologues, isolated homoeologous loci from selective competition. Following mutation at one locus the original function was continued at a homoeologous locus, without one allele replacing the other through the action of selection as would have happened if there had been intergenome pairing and polysomic inheritance.

Nevertheless, while this aspect of the operation of the mechanism is of considerable importance, the fundamental significance of the chromosome V mutant has undoubtedly been its influence on fertility. High fertility and productivity, and constancy of the genotype, are the prerequisites of a crop grown for seed, and it was probably by the action of selection on these characters that the mutant was originally fixed in the tetraploid population. Thus it can be asserted without exaggeration that the development of the wheat crop has depended upon the chromosome V mutant—without this or a comparable change, polyploid wheat would never have achieved any agricultural significance. Indeed, without the prior occurrence of a meiotically diploidized tetraploid, the hexaploid condition would probably never have been attained.

Finally it is interesting to reflect upon the impact of these evolutionary changes in wheat, on wider issues. The formation of the first settled communities in the Near East depended upon the establishment of agriculture. The earliest civilizations, which eventually developed via Greece and Rome into what is known as Western civilization, grew from such communities. Apparently the primary crops of the first agriculturalists were wheat and barley—in effect agriculture was founded on these crops. To some degree, therefore, the emergence of our culture may be seen to be linked to a mutation which affected the meiotic chromosome pairing, and hence the fertility, of a wild grass.

VI

THE HISTORY AND RELATIONSHIPS OF CULTIVATED POTATOES

by KENNETH S. DODDS

THE genus *Solanum* has a world-wide distribution but its tuber-bearing species are only found in the New World. They fall into two clear groups—wild species and cultivars. The potato was brought to the Old World after the Spanish Conquest in the sixteenth century but there is no tangible evidence of how it was domesticated by South American man. Ancient cultures of South America had no writing and the potato itself is not the sort of object to be fossilized.

The development of modern botanical concepts of the potato and its relatives can be said to have started with the Russian plant collecting expeditions to South America in the years between 1925 and 1932. These expeditions stimulated world-wide interest in the potentialities of the indigenous potato varieties of the Andes for potato breeding and also in wild species of tuber-bearing *Solanums*. In due course, Sweden, Germany, the United States of America and the United Kingdom sent their own expeditions and as a result there is now a great wealth of potato material at many agricultural research centres outside South America, which is being used for cytogenetical investigations and in breeding programmes. But if we are to unravel the evolutionary history of this major crop, there is a very real need for laboratory studies to be supported much more extensively by field work in South America. We now have a fair picture of the differentiation of the cultivars but our knowledge of the distribution, ecology and variability of the wild tuber-bearing species is really no more than fragmentary.

A large assortment of wild species is distributed more or less continuously from the south of the United States to southern Chile. Some species are relatively constant and seem to have entered an evolutionary cul-de-sac whereas others are very variable. They are found at many altitudes in the temperate latitudes and occupy many different habitats, in arable areas, in woodlands and in deserts. In tropical latitudes, they do not occur below about 6000 ft.

The vicissitudes to which potato taxonomy is subjected can be judged from the diverse opinions on classification mentioned in the recent reviews of potato cytology and genetics by Swaminathan and Howard (1953) and Howard (1960). Much of the present confusion has arisen because in the past taxonomic revision has been attempted with very little knowledge of the genetical architecture of the wild plants. Species have been distinguished and named in both wild forms and cultivars without any recognition of the intrinsic differences between their patterns of differentiation. Usually, unrepresentative plant collections have been studied and, almost invariably, field observations have been too few. Fortunately, it may be argued that many of the wild species have no direct evolutionary connexion with the cultivars so that the precise way in which they are joined to the taxonomic skeleton of the genus is not pertinent to the present discussion.

Tuber-bearing *Solanums* are classified in the sub-section Hyperbasarthrum of the genus (section Tuberarium; sub-genus Pachystemonum). The basic chromosome number is $x = 12$ and there are diploids, triploids, tetraploids and hexaploids. Hawkes (1956*b*) recognizes seventeen series but Correll (1962)* has recently increased the number to twenty six. Two of these series contain species which do not form tubers and I shall not mention them further. Of the remaining twenty four, ten contain species which occur in Mexico and Central America and thirteen are concerned with South American species. Only one series has species with major distributions in both continents (Table 5).

Species of the five series Bulbocastana, Cardiophylla, Clara, Pinnatisecta and Trifida are chiefly Mexican, although isolated communities of *S. bulbocastanum* are found in central Colombia. All the species have white flowers with stellate corollas and they are diploids; however, sterile triploid clones do occur in three of them. Recently Pandey (1960) and Malheiros-Gardé (1959)

* This manuscript was revised in the press to conform with Correll's systematic treatment of section Tuberarium.

have suggested that some of these species have a two-locus system of incompatibility and not the one-locus system that is found in the species of South America. This suggests that a major divergence in the genetic control of the breeding system has accompanied geographical isolation.

TABLE 5. *Number and ploidy of species within series*

Region	Series	Number of species with					
		24	36	48	60	72	?
North and Central America	Borealia			1			2
North and Central America	Cardiophylla	2	(2)				
North and Central America	Clara	1					
North and Central America	Demissa	1				4	1
North and Central America	Longipedicellata		1†	5			1
North and Central America	Morelliformia	1					
North and Central America	Pinnatisecta	6†					2
North and Central America	Polyadenia	1					
North and Central America	Trifida	1					
North and Central America / South America	Bulbocastana	1	(1)*				1
North and Central America / South America	Conicibaccata	2		6		1	11
South America	Acaulia			1		1†	
South America	Circaeifolia	2					
South America	Commersoniana	2	(1)				1
South America	Cuneoalata	2					
South America	Ingaefolia			1			4
South America	Megistacroloba	4					7
South America	Minutifolia						2
South America	Piurana	4					11
South America	Tarijensa	2					
South America	Transaequatorialia	18					26
South America	Vaviloviana	1‡					
South America	Yungasensa	1					3
South America	Tuberosa: Wild	2	(2)				2
South America	Tuberosa: Cultivars	1	1†	1	1†		

* Clone of diploid species.
† Includes a probable natural interseries hybrid.
‡ Possibly a tetraploid form also.

The series Morelliformia and Polyadenia each contain only a single species. *S. morelliforme* is a self-fertile epiphyte with simple leaves and small white stellate flowers. *S. polyadenium* is a very characteristic diploid species found in Central Mexico. The plants are covered with glandular hairs and have an objectionable smell; the white flowers are self-fertile and they have substellate corollas. These species do not hybridize with other species and they can be assumed to have played no part in the main evolution of the genus.

The chromosome number of only one of the three species included in series Borealia has been determined ($2n = 48$). Its species are distinguished from all those in other series found in middle and North America by the three-leaflet aspect of the leaves and the rotate corollas.

Series Demissa contains a rather heterogeneous group of Mexican tuber-bearing species characterized chiefly by the high pedicel articulation of the flowers and the rotate corollas with very short lobes. One of its species is a diploid, the chromosome number of another has not yet been determined, and four species are hexaploids. A well-known member is *S. demissum* ($2n = 72$) used in potato-breeding programmes as a source of the major (R) genes controlling disease caused by *Phytophthora infestans*. Some of these alleles have been identified in the very variable self-fertile tetraploid species *S. stoloniferum* which, together with four other tetraploid species, is classified in the series Longi-pedicellata.

The series Conicibaccata is so named because of the ovate-conical berries that are a characteristic feature of its members. It contains twenty species of which two are diploids, six are tetraploids, one is a hexaploid and the chromosome numbers of the remaining eleven have not been determined. For the most part they are found in wet mountain forests. *S. agrimonifolium*, *S. oxycarpum* and *S. Woodsonii*, all tetraploids, and *S. reconditum* are central American species whereas the others are confined to South America. Series Piurana and Ingaefolia are closely allied, both containing species that occur mainly in southern Ecuador and northern Peru. The former contains a group of vegetatively rather distinct species having leaflets that are either glabrous or glabrescent and verrucose on the upper surface or thick and subcoriaceus; the few plants of *S. tuquerrense* that have been studied, a tetraploid species in series Ingaefolia, were all self-sterile. No experimental study has been made of any of the species in the two small series Minutifolia and Yungasensa. *S. Wittmackii* ($2n = 24$) series Vaviloviana is endemic to the lower vegetation zone of coastal Peru. The plants have a glandular pubescence like those of series Tarijensa whose species occur from central Bolivia to north-west Argentina, with one species, *S. polyadenium*, in Mexico. Series Circaeifolia contains two diploid Bolivian species, *S. circaeifolium* and *S. capsicibaccatum*.

Series Commersoniana contains three species, two of which *S. chacoense* and *S. commersonii*, are aggressive, extremely polymorphic and extend over a wide area in South America, being found in north and central Argentine, Paraguay, Uruguay, and south Brazil. Triploid clones of both these species occur. Correll (1962) writes 'the series Commersoniana, of South America, and Pinnatisecta, of north and central America, are here maintained separately, primarily, if not solely, on the basis of the apparent geographical isolation of the species which comprise each series'.

The series Cuneoalata contains two diploid species *S. infundibuliforme* and *S. zerophyllum*. It may be questioned whether this series should be separated from series Megistacroloba which contains four diploid species and seven species whose chromosome numbers are not known. The members of both species are small straggling plants which have so far been collected in central Bolivia and north-west Argentine and are characterized chiefly by the development to a variable extent of wedge-shaped wings between the leaflets and along the rachis of the leaf. An intensive experimental field study of this group of species is needed as the basis for a re-appraisal of their validity.

Series Acaulia contains only one species, *S. acaule* ($2n = 48$). It is noteworthy because it figures in the pedigree of some of the cultivated potatoes (Bukasov, 1939), a point to which I shall refer below. *S. acaule* is a frost resistant species with a rosette habit and it is very prevalent above about 12000 feet on the altiplanos of Central Peru, Bolivia and the north-west Argentine. Although the species is very self-fertile, often the plants are so heavily grazed that they do not get a chance to set fruits.

The large series Transaequatorialia contains more than forty species. It is not easy to define the characteristics which are used to place a species in this series and, indeed, there is some justification for the conclusion that it receives either deliberately or unwittingly all the species that cannot be classified in any of the other series with more obvious botanical distinctions. The differences between two of its species may be as great as those between species from two other series. Species of Transaequatorialia are found at high and low altitudes, among natural vegetation and as weeds. They may be upright or straggling plants and have simple or highly dissected leaves with or without interjected leaflets and a glandular pubescence. The flowers may be white or coloured and the corolla vary in shape from rotate to semi-stellate. All those which have been examined are self-incompatible diploids.

The last series, Tuberosa, contains two littoral species, two montane species and cultivated potatoes. It illustrates the limitations in a phylogenetic sense of a serial arrangement of species, inasmuch as Transaequatorialia contains nearly all the species that have been named as putative ancestors of the cultivated potato.

In general meiosis in hybrids between diploid species is regular and little evidence has been found of genomic differentiation; that is, genetic divergence is not reflected in chromosome structure. Most of the triploids that occur are probably autotriploids the origin of which can be attributed to the functioning of unreduced gametes within the species. These triploids have survived by virtue of their capacity for vegetative propagation. Marks (1958) has shown, however, that the hybrid triploid *S.* × *vallis-mexici* is very similar in appearance and cytological behaviour to the triploid hybrid obtained by crossing two species with which it is sympatric, namely *S. verrucosum*, a diploid species of series Demissa, and *S. stoloniferum*, a tetraploid of the species Longipedicellata. This hybrid species and the experimental triploid when doubled by colchicine treatment gave hexaploids which crossed very easily with all the hexaploid Demissa species except *S. guerreroense*; but they did not resemble any of them morphologically. Nevertheless, it seems highly probable that the hexaploid species of Demissa have evolved by doubling of triploid hybrids of *S. verrucosum*, *S. stoloniferum* and their near allies. This view is supported by the observation that certain *R* genes controlling a hyper-sensitivity reaction to *Phytophthora infestans* are common to *S. stoloniferum* and *S. demissum* (Schick, Schick and Haussdörfer, 1958; McKee, 1962).

The tetraploid wild species usually form twenty-four bivalents at first metaphase and in consequence are thought to be allotetraploids and the hexaploids form thirty-six bivalents and appear to be allohexaploids. The F_1 hybrids from crosses between the hexaploids form mostly bivalents and univalents, a behaviour which is not at variance with a supposition that some genomic differentiation has occurred in these species (Marks, 1955). However, in view of the demonstration of genetic diploidization in some polyploids of wheat (Riley, 1960), it is now an open question whether a similar phenomenon can be detected in these polyploids.

The indigenous cultivated potatoes of the New World have a narrower distribution than the wild species. The main centre of diversity is in South America, between about 10°N and 25°S of the Equator. Moreover, as they require a temperate climate, they are confined in these latitudes to the Andes, at altitudes above about 6500 ft. Some clones occur on the coastal lowlands in southern Chile, however, between about 40° and 45°S but they are probably relatively recent introductions to the area.

The maximal variability of the cultivars is on the altiplano around Lake Titicaca in Bolivia. Most of the potatoes grown in this region are tetraploids, but diploids, triploids and pentaploids also occur. Usually it is not possible to distinguish the different levels of ploidy except by chromosome counts for all these potatoes are part of the same vast spectrum of variability. Indeed, in Bolivia, plants of different ploidy are usually grown together indiscriminately. Tubers from mixed diploids and polyploids are harvested together and often sold still mixed in the local markets. The cultivation of such mixtures is not as prevalent elsewhere in the Andes, however; for example in Ecuador, Colombia and Venezuela at the northern end of the range, the diploids are grown separately from the tetraploids. The reason for this is that the diploids provide a series of early maturing clones with non-dormant tubers which are well adapted to virtually continuous growth in the milder environments of the warm valleys off the altiplano proper. They are very commonly planted around the houses or on the less accessible hillsides in situations which are relatively free from frost and not as exposed as those occupied by their counterparts in Bolivia. In the northern zone, the diploids are the so-called Chaucha (early) potatoes. They are said to have a good flavour and are used particularly in soups and stews. Nevertheless, the bulk of potato production in these countries comes from tetraploids, which are grown in the fertile valleys between say 6500 and 9000 ft.

In summary, then, the Andean cultivars are basically a great complex of diploid, autotriploid and autotetraploid clones, the three categories covering essentially similar spectra of variability and being often intermingled in the field. I shall now consider the origin of this complex and deal also with a small group of polyploid interspecific hybrid clones which are in the evolutionary sense peripheral to the main mass.

First consider the hybrids just mentioned: they consist of a small sector of the triploids and all the pentaploids. These particular triploids are frost resistant and are thought to have been derived by hybridization on the Bolivian altiplano between *S. acaule* ($2n = 48$) and cultivated diploids ($2n = 24$) (Bukasov, 1939; Hawkes, 1956*b*). They have a semi-rosette habit and leaves that are borne somewhat erectly, with obtuse leaflets, being very similar to those of the rosette-forming *S. acaule*. The pentaploids do not look as much like *S. acaule* as do the triploids but they have a semi-rosette habit, rather straight stiff leaves and high pedicel articulation. They are believed to be derived by natural crossing between the hybrid triploids ($2n = 36$) and cultivated tetraploids ($2n = 48$) (Hawkes, 1956*b*). These hybrids are not very important agronomically. Some of the pentaploids, however, are fertile, more so indeed than many of the autotetraploids, and they could be a means of introgression of *acaule* genes into the general population as Simmonds (personal communication) has pointed out.

There has been considerable speculation in the literature about the rôle that wild tuber-bearing species have played in the evolution of the cultivars but virtually no evidence apart from comparative taxonomy has been adduced to support the many conjectures that have been put forward. The putative parentages of the two hybrid polyploids that have just been discussed can be accepted with some confidence; *S. acaule* is so distinct that the only way of getting the particular combination of characters possessed by the hybrids would seem to be by a cross in which it was one of the parents. But other speculations of ancestry based entirely on morphological similarities are more open to criticism; for example, as the cultivars offer so much variability from which to choose, some of them can be matched closely in morphology with one or other of the wild species of series Transaequatorialia. This sort of comparison only gives an illusion of progress unless experimental evidence is also provided further to substantiate the evolutionary speculations.

A first step towards an understanding of the evolution of potatoes would be taken if we could define the differences between a cultivated and a wild plant. The haulms of some diploid species and cultivars may look extremely alike but this does not mean that they are biochemically similar; for example, Harborne and Corner (1961) find that the 3-glucoside of caffeic acid that occurs in the berries of wild species is replaced in cultivars by umbelliferone and *p*-coumarylglucose. Tubers have been the organs subjected directly to man's selection and one would expect their differences to be accentuated. It is surprisingly difficult, however, to describe the basic differences between tubers from cultivated and wild plants. The tubers of cultivars tend to be larger than those of wild species but this is not an intrinsic difference and sometimes may be no more than a reflection of greater photosynthetic activity dependent on plant size. Cultivars have more kinds and patterns of anthocyanins than wild species but an increase in the variability of pigments is commonly associated with the domestication of crop plants. The basic anthocyanin system of the wild species can still be identified in the cultivars though it does seem to have been considerably elaborated under domestication. Thus it seems reasonable to assume that the Mendelian variability of the pigment system has arisen since domestication probably in response to the interest that early South American man had in bright colours and clear patterns; an interest that is still evident in South American Indians. Cultivars have a lower concentration of alkaloids in the flesh of their tubers than wild species but these biochemical differences have not been studied in any detail (Schreiber, 1954). It does seem likely, however, that the loss of bitter principles such as alkaloids was a major selective event in the domestication of the potato (Cárdenas, 1956). Correll (1948) mentions the crisp, astringently bitter tubers of *S. bulbocastanum* and the tender, mildly sweet tubers of some strains of *S. demissum* and *S. longipedicellatum*. He was told that, in some states of Mexico, the Indians bring wild tubers to the local markets at various times during the year and that they are frequently used in soups and stews. There is no record of this being done either now or in the past by the inhabitants of South America. At present we cannot go far beyond the position of realizing that the realm of the Incas and their antecedents was a great centre of plant domestication. Maize and potatoes were the staples of the sierra and they were supplemented by the cereal-like crop quinoa (*Chenopodium quinoa* Willd.) and by three other edible tuber plants whose cultivation today is being gradually abandoned. They are añu (*Tropaeolum tuberosum* Ruiz & Pav.), oca (*Oxalis tuberosa* Molina) and ulluco (*Ullucus tuberosum* Caldas). These were the basic subsistence plants of the Andes and their joint study might be expected to help in unravelling the botanical mysteries of each. It is interesting to note that bitter varieties of oca must be mellowed by several days' exposure to sunshine before being eaten. Like potatoes, they can be converted into a desiccated form known as 'chuño'.

Seemingly wild potatoes that are often found as weeds in cultivated fields or flourishing on waste land have been interpreted both as escapes from cultivation and as species part way along the road to domestication; for example, Perlova (1958), according to Howard (1960), implicates *S. herrerae*, *S. leptostigma* and *S. molinae* in the lineage of cultivated tetraploids but the material which was described under these binomials is now included in *S. tuberosum* L. by Hawkes (1956*b*). This latter author believes that '*S. tuberosum* arose either directly from an ancestral form of *S. stenotomum* by a process of simple chromosome doubling, or as a spontaneous amphidiploid hybrid between the more ancient diploid cultigen *S. stenotomum* and the diploid wild species *S. sparsipilum*' (Hawkes, 1956*a*). *S. sparsipilum* is a very polymorphic species which is widely distributed from central Peru to Bolivia; vegetatively, its members look very much like some cultivated potatoes. Diploid cultivars of the Stenotomum Group were 'no doubt derived from wild ancestors related to the present day *S. leptophyes*, *S. soukupii*, *S. canasense* etc.' (Hawkes, 1958), other diploid species in the series Transaequatorialia. Brücher (1958) chooses *S. vernei* ($2n = 24$) as the ancestral species of cultivars but in Hawkes's view 'It is doubtful whether *S. vernei* has played any part in the formation of cultivated potato species'. Cárdenas (1956) also favours *S. vernei* and possibly species of the series Commersoniana. These examples illustrate the liberal sentiments that are evoked by this problem but it must be clearly understood that no experimental evidence has been published to support the claims that these species are the precursors of our cultivated potatoes. Even if they are good species, little can be said about their evolutionary significance until they have been studied cytogenetically. The difficulties that stand in the way of this are manifold. With little, if any, genomic differentiation between species, metaphase cytology is unlikely to be particularly helpful. Gottschalk (1954) and Gottschalk and Peters (1954) have studied the pachytene stage of meiosis and have paid particular attention to the distribution and number of chromomeres in the heterochromatic segments of the chromosomes of different species but, unfortunately, their results are too controversial to be regarded as a useful contribution (Wangenheim, Fransden and Ross, 1957; Howard, 1960). Fiedler and Schreiter (1959) have made and analysed a few excellent preparations of pachytene stages. If this can be done on a more extensive scale, further progress should be possible in the study of the comparative structure of *Solanum* chromosomes.

The customary way of examining the relationships between species is by studying the interactions of their genetic systems in hybrid progenies. In effect, what we want to know about the potatoes are the genetic changes that have occurred during domestication. It is galling to realize how little we actually know about the physiological genetics of characters that are of significance in the domestication of man's crop plants. Now that the broad principles of Mendelian and quantitative genetics are well established, it is to be hoped that some of these less tractable problems in the physiology and genetics of higher plants will become more fashionable as topics for research.

Comparative taxonomy and, when it is possible, experimental hybridization are extremely valuable methods of studying general relationships between species. Once these have been established, however, further progress is made only if specific genes can be identified and used as evolutionary markers. It has occurred to me that the *S* alleles of the incompatibility system are ideally suited for this purpose in *Solanum* and so I am approaching the problem of the origin of cultivated potatoes by making a comparative study of the *S* alleles in diploid wild species and cultivars. There would seem to be no *a priori* reason why the specificities involved in the style-pollen tube reaction should be modified during the process of domestication if sexual fertility has remained unaltered and the mating system is still determined by a single locus. In the tests so far carried out, about twelve wild species and four cultivars have been used; no style-pollen tube reactions that would indicate the occurrence of identical *S* alleles have yet been found.

It is axiomatic that man has created genetic divergence in his crop plants by selecting types adapted to his needs. In potatoes, this would first occur at the diploid level and differentiation based on spontaneous triploids and tetraploids would take place later. Hawkes (1956*b*) recognizes four species of cultivated diploid potatoes. In families of these four 'species', Dodds and Paxman (1962) measured seven characters: leaf dissection, density of leaf hairs (two characters used as taxonomic criteria), leaflet index, pedicel abscission, length of calyx, flowering time and tuber production. Two of the 'species' differed significantly in two characters (flowering time and leaf dissection) but phenotypic segregation of all the characters was so wide that even when a conservative estimate of the range of segregation was used (family mean $\pm$ standard deviation), there was still a wide overlap between 'species'.

The diploids are interfertile and meiosis in F_1 hybrids between clones from extremes of the geographical range is regular. They are outbreeders with a gametophytic system of incompatibility controlled by many alleles at a single *S* locus. Dodds and Paxman (1962) have demonstrated the potential panmixis of the diploids by showing that some of their *S* alleles are distributed throughout the geographical range of the cultivars. Two alleles of an Ecuadorean clone proved to be of particular interest. Although they were identified in diploid clones over the whole geographical range, they were very common in Colombia and

Ecuador, 72·5 per cent (74/102) and 61·9 per cent (26/42) of the clones from these two countries having one or other of the alleles, and comparatively rare in Peru and Bolivia with frequencies of 25 per cent (7/28) and 15 per cent (6/40), respectively. Recently the reverse situation has been observed; one of the alleles of a clone from Cochabamba is very common in Bolivia and has been detected throughout the geographical range of diploids, getting progressively scarcer northwards to Venezuela. A preponderance of certain *S* alleles would be expected to occur if the *S* locus was linked to another locus with agricultural significance, for example, dormancy, or if the similar groups were evolved each from a small nucleus of original material. Investigations are being carried out to try and solve this problem.

TABLE 6. S *genotypes of twenty randomly selected Bolivian clones*

	S alleles												
	3	13	14	15	16	17	18	19	20	21	22	23	24
3		2		1			1	1	1		1	1	
13			3		1			1		1			
14					1	1		1					
15					1								
16													
17													1
18													
19									1				

Another feature of the breeding system of the diploids is shown in Table 6 which gives the *S* genotypes of a random selection of twenty Bolivian clones. Only seventeen of the forty alleles are different. As the assumptions that are necessary to apply the mathematical models used by Bateman (1947) and Raper, Krongelb and Baxter (1958) with populations of clover and *Schizophyllum*, respectively cannot be made in potatoes, no estimate can be given of the total number of different *S* alleles in the diploid population. Different genotypes are not equally represented in the population because, clearly, some clones are more favoured by the growers than others.

Dodds and Paxman (1962) were able to show not only that the breeding system of cultivated diploids is based on a common series of *S* alleles but that the genetic control of anthocyanin development is also similar throughout the group. Thus, bearing in mind the interfertility of the diploids, their opportunities for crossing by the agency of bees, their continuous range of variability and their common breeding system, it must be concluded that they are, in fact, one large, potentially panmictic, population in which differentiation has not reached the level of speciation.

It is quite evident from looking at diploids as they now exist in South America, that certain components of the potato population have been selected and raised almost to varietal level. The different polyploids are grown mixed together on the Bolivian altiplano and they must all be assumed to have the same general agronomic and culinary characteristics. In southern Peru, however, a yellow fleshed diploid called Papa amarilla, which is renowned for its flavour, is grown separately specially for the markets of Lima. Dodds and Paxman (1962) have shown that this white flowered cultivar is a single clone. Again, diploids of the northern complex are quite distinct. They mature early and their tubers are not dormant. Presumably, potatoes with these growth characteristics have been selected to enable the gentler climate of the Andean valleys to be fully exploited. With relative freedom from frost, two crops a year can be grown, or even three when irrigation is used. Although the northern diploids are similar to each other in maturity, they vary in many other characters. This variability may have arisen and been retained by some form of assortative mating; for example, possibly by crossing between tuber-plants and volunteer seedlings. It cannot be claimed that the north Andean peasants have yet reached with diploids what is usually regarded as the highest level of management in vegetatively propagated crops; namely, the establishment of clonal plantings. But they are not far from it in some areas. Certain groups of potatoes with local names such as Amarilla, Manzana, Negra, etc. are fairly distinct and each consists of no more than a few clones. These diploids are valued for soups and stews and there is a firm demand for them in the city markets. This will probably ensure that some clones are kept in cultivation even though, on the whole, the cultivation of diploids in the northern region is being abandoned in favour of the higher yielding tetraploids.

Some indication of the relative proportions of diploids, triploids and tetraploids in the potato population at the centre of variability is given in Table 7. The chromosome numbers determined by Ochoa (1958) were obtained from a sample of 300 clones collected in the basin of Lake Titicaca in Bolivia. Fairly similar proportions are shown in the samples obtained by Dodds and Paxman (Unpub.) from Peru and Bolivia even though these investigators tried, by the appearance of tubers, to collect only diploids. The results show that this can only be done efficiently in the northern Andes where the diploids have non-dormant tubers and can be identified in consequence, as mentioned above. At the main centre of variability, however, this situation does not hold. Attention may be focused on the resemblances between tubers of different ploidy by the observation that all three or any two levels of ploidy were found to occur

TABLE 7. *Relative frequencies of diploids and polyploids in peasant cultivations*

		$2x$	$3x$	$4x$
Dodds and Paxman	Venezuela	4		2
	Colombia	106	4	1
	Ecuador	37	4	29
	Peru	32	25	84
	Bolivia	44	9	44
Ochoa	Bolivia	159	65	176

in separate samples from twelve stalls in different markets in south Peru and Bolivia. Vegetative characters are not much better as a means of distinguishing plants of different ploidy. In a test at Bayfordbury, using a mixture of twenty diploids and twenty tetraploids planted randomly, four observers familiar with potatoes tried to classify the plants according to ploidy, by haulm characters. All did better than chance (0·5) but only three were significantly better; two at the 1 per cent and the third at the 5 per cent level. Some slender diploids with narrow leaflets can be identified but triploids and tetraploids in particular look very much the same. In general, the indigenous potatoes in south Peru and Bolivia are more straggling in habit and have more delicate foliage than European potatoes; they are also adapted to short days.

Tetraploids are thought to be autotetraploids, and that they are self-fertile agrees with this supposition. Presumably the gametophytic incompatibility system that is functional in diploids breaks down in tetraploids as a result of competitive interaction between alleles in diploid pollen. Nevertheless, tests show that polyploids appear to retain the specificities of the *S* alleles in the styles. Although the tetraploids are self-fertile, they very rarely set berries by selfing in the undisturbed surroundings of an insect-proof glasshouse; they must be hand pollinated to do so. In nature, it seems likely that the plants are shaken sufficiently to ensure some self-pollination but, as the flowers are worked by solitary bees, it is also highly probable that a good deal of outcrossing occurs.

Unreduced gametes are not uncommon in tuberous *Solanums* and so spontaneous polyploids might be expected to occur not infrequently. But once tetraploids had been established, the greatest divergence would be given by crossing within and between diploids and tetraploids. For example, if we except the small group of triploids thought to be derived by hybridization with *S. acaule*, it is reasonable to suppose that nearly all the others have arisen by intercrossing between cultivated diploids and tetraploids. General intercrossing of this kind in nature would seem to be the only practicable way of arriving at the population of diploids, triploids and tetraploids that is so characteristic of the Bolivian altiplano (Table 7). Proof that this was the sequence of events is being sought by studying experimental crosses between diploids, triploids and tetraploids and by examining the distribution of certain *S* alleles in the natural population of cultivars. It is interesting in this connexion to note that tetraploids are less fertile than triploids in experimental crosses with diploids as male parents. But whereas all the progeny of triploids are approximately diploid, the tetraploids give diploids, triploids and tetraploids. In the $4x \times 2x$ cross, triploids are quite frequent but unreduced male gametes are strongly selected to yield many tetraploids; the diploids, which are fairly frequent, are polyhaploids (Table 8). These are known to be easily obtainable from the European potato by suitable matings (Hougas, Peloquin and Ross, 1958). The diploid male parent used in the crosses recorded in Table 8 had the genotype $S_{11}S_{12}$ and was homozygous for coloured flowers (*PpRRFF*; Dodds and Long, 1955, 1956) but none of the polyhaploids had either of these two *S* alleles and all had either white or flecked flowers—proof of their origin from unfertilized embryo-sacs. So far, three alleles that are common in diploids have been identified in both triploids and tetraploids; a finding which can be accepted as proof of the common origin of the cultivars.

The bulk of potato production in the Andes is from tetraploids grown on haciendas in the valleys and the peasants in the northern region are turning more and more to tetraploids as they become familiar with their potentialities for higher yield. This observation suggests that tetraploidy was favoured originally because of its beneficial effects on yield but this cannot be taken for granted by any means. Peloquin and Hougas (1960) found that most of a group of twenty-nine polyhaploids were smaller

TABLE 8. *Progeny from triploids and tetraploids crossed with C.P.C. 979* (2x = *24*)

		Progeny		
	Seeds per berry	$2x$	$3x$	$4x$
$3x \times 2x$	$10{\cdot}0 \pm 1{\cdot}73$	28*	0	0
$4x \times 2x$	$1{\cdot}9 \pm 0{\cdot}19$	8	24	26

* Includes four with $2x = 25$ and one with $2x = 26$.

and yielded less than the parental tetraploids but one or two exceeded the parents in vigour and yield. The effects of ploidy are not separable in their data from genetical effects, and the haploids were derived from highly selected North American varieties (mainly Katahdin). But even with these limitations, the relatively high frequency of good haploids fosters the suspicion that the selective value of tetraploidy in indigenous potatoes might prove not to be purely and simply higher yield. Satisfactory comparisons from which to judge the evolutionary significance of tetraploidy would have to be made in the Andes and before this could be done, diploids, triploids and tetraploids would have to be sorted out; and, because most indigenous clones are infected with virus, the tests would most probably have to be done with seedlings. These are just a few of the many obstacles that would have to be overcome in order to get more information on this problem. As I have said earlier, diploids, triploids and tetraploids are often grown together in Bolivia, from which it may be argued that no particular category of ploidy is outstandingly superior on the altiplano. It may be suggested that tetraploids gain their superiority at somewhat lower altitudes in the more fertile Andean valleys where they are grown exclusively. We have no knowledge whatever of the general physiology of the potato plant in these two environments, however, and so there are no facts on which to base further discussion. Comparative field trials at both kinds of site are very much needed.

History suggests that potatoes were first brought to Spain in the sixteenth century, whence they spread throughout Europe, and there is good reason to believe that a second introduction was made independently at about this time to England (Salaman, 1949). In the British Isles, the potato was first grown in quantity in Ireland and then spread slowly to other parts. Cultivation was fairly widespread by the early eighteenth century and the number of varieties was on the increase as a result of seedling selection.

Juzepczuk and Bukasov (1929) described potatoes from southern Chile not unlike our present European varieties and concluded that the potatoes first introduced to Europe came from this part of South America. From topographical considerations alone, it seems very unlikely that southern Chile was a centre of potato differentiation. Darwin (1845) described Chiloé as being densely wooded except for small coastal clearings. Much of the forest both on Chiloé and the adjacent mainland is still standing today. Salaman (1954) found a ratio of about 4:1 Andean to European-like potatoes among Chilean varieties, as scored by 'leaf index', a measure of the open disposition of leaflets. A recent collection of about 150 such varieties was judged by inspection of growing plants to contain Andean, European-like and intermediates roughly in the proportion 3:3:2 (Simmonds, unpublished). It is by no means inconceivable that the European-like varieties that occur in Chile are the remnants of forgotten eighteenth-century introductions. Salaman and Hawkes (1949) have pointed out that the old drawings of European potatoes show plants which look like present day Andean varieties and they suggest that, in the sixteenth century, the potato would be much more likely to survive the journey to

Europe from the northern Andes than from southern Chile. These and botanical considerations make them favour the view that the potato was introduced to Europe from the northern Andes.

The first European potatoes, as the herbals show, must have looked very similar to the present day Andean tetraploids. But, in South America, potatoes have changed very little in the past two or three centuries whereas in Europe they were soon modified by selection, ultimately to become our modern varieties. Indeed, by the time Linnaeus (1753) named the potato *Solanum tuberosum* L., some of the varieties (including the Linnaean lectotype) had distinctly higher leaf indices than Andean potatoes (Salaman and Hawkes, 1949). Recently, Simmonds (unpublished) has re-examined the data on leaf indices of seventeenth-century clones and modern European clones published by Salaman and Hawkes (1949) and by Salaman (1954), respectively. He concludes that the Linnaean lectotype was taken from a population of varieties, now extinct, that was transitional between modern Andean and European clones. That this particular level of differentiation supplied the lectotype can now be regarded as accentuating the appropriateness of the binomial *Solanum tuberosum* for all cultivated potatoes, regardless of origin and ploidy (Dodds, 1962). Modern varieties have been selected without changing the capacity of all fertile cultivars to interbreed. Except for the triploid hybrid *S.* × *juzepczukii*, diploids, triploids and tetraploids whether European or Andean have the same basic heritage. It is a strange quirk of history that today the variability of these groups is being thoroughly intermingled in the Andes by programmes of hybridization between the indigenous varieties and varieties introduced from Europe and North America; potato breeders in South America are busily trying 'to make the best of both worlds' in potato production.

VII

THE EVOLUTION OF FORAGE GRASSES AND LEGUMES

by J. P. COOPER

MOST food crops have been cultivated for centuries, and in many cases appear to have arisen under cultivation. While their patterns of distribution and centres of diversity are well known, it is often difficult to trace their origin from wild ancestors in spite of combined taxonomic, cytogenetic and archaeological investigations.

In forage plants on the other hand, while a number of grass and clover species are sown extensively in cultivated pastures, many of the grazing animals of the world are supported on natural grasslands and the distinction between wild and cultivated species is not so clear cut. Furthermore, the deliberate cultivation of forage crops has developed in comparatively recent historic times.

The forage grasses and legumes thus present a group which is in active process of domestication, and can therefore provide valuable information on the genetic changes occurring under human selection. These changes are now proceeding at an increasing rate as a result of:

(i) increasing opportunities for plant migration and for deliberate plant introduction,
(ii) extensive modifications of the environment resulting from changes in farming systems such as large scale mechanization and irrigation,
(iii) the purposeful selection and hybridization of the plant breeder.

The evolution of any crop plant poses two interrelated problems; how did its present pattern of genetic variation originate, and what are its potentialities for future change? This last problem is of immediate interest to the plant breeder. The present pattern of variation in a crop can be conveniently considered at two levels; firstly the macro-evolutionary relationships between species and genera, taxa between which gene exchange is difficult or impossible; and secondly the micro-evolutionary pattern of variation *within* the species, where free exchange of genes can occur.

One of the most important sources of evidence on the past history of any taxonomic group is its present pattern of distribution. In the case of most food crops, it has been possible to identify distinct centres of diversity, usually associated with the early cultivation of the crop. In the forage crops, however, this association is not so clear and a distinction must be made between the distribution of wild and cultivated species.

The world distribution of the more important of the tribes of the Gramineae in relation to their past history and present climatic adaptation has been studied by Hartley (1950, 1958*a*, *b*). He finds that certain tribes, such as the Agrosteae and Festuceae, are widely distributed in all hemispheres, but predominate in temperate latitudes or at high altitudes in the tropics. The winter isotherm of 50°F for the coldest month delimits the regions where these tribes are found in super-normal frequencies, and rainfall seems to have no effect on their distribution. They show no obvious centres of diversity and Hartley concludes that these tribes have had a long evolutionary history and have reached their maximum possible range, being limited only by climatic factors.

The Andropogoneae and the Paniceae, on the other hand, have a predominantly tropical distribution, high winter temperature being an important determining factor. The Andropogoneae are favoured by a high midsummer rainfall, i.e. by a monsoon climate, and show a main centre of species differentiation in South-east Asia (as shown in Fig. 25) with rather fewer species in America than would be expected on climatic grounds. The Paniceae (Fig. 26), on the other hand, have a predominantly New World distribution, with an apparent maximum density of species in the West Indies and tropical South America. They seem to be favoured by a high total annual rainfall as occurs in the equatorial long wet season. More detailed studies, however, indicate that the greatest diversity at the *generic* level in this

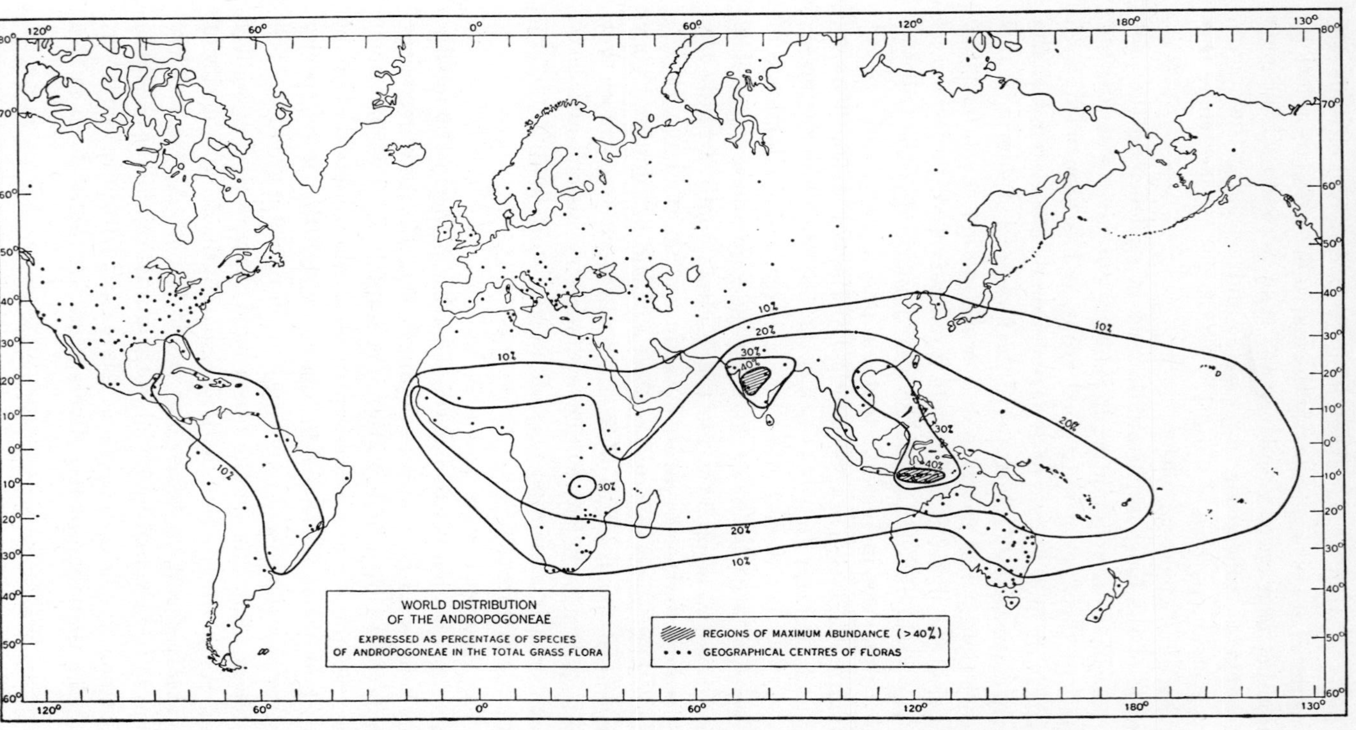

Fig. 25. Map of world distribution of the Andropogoneae. (From Hartley, 1958*a*.)

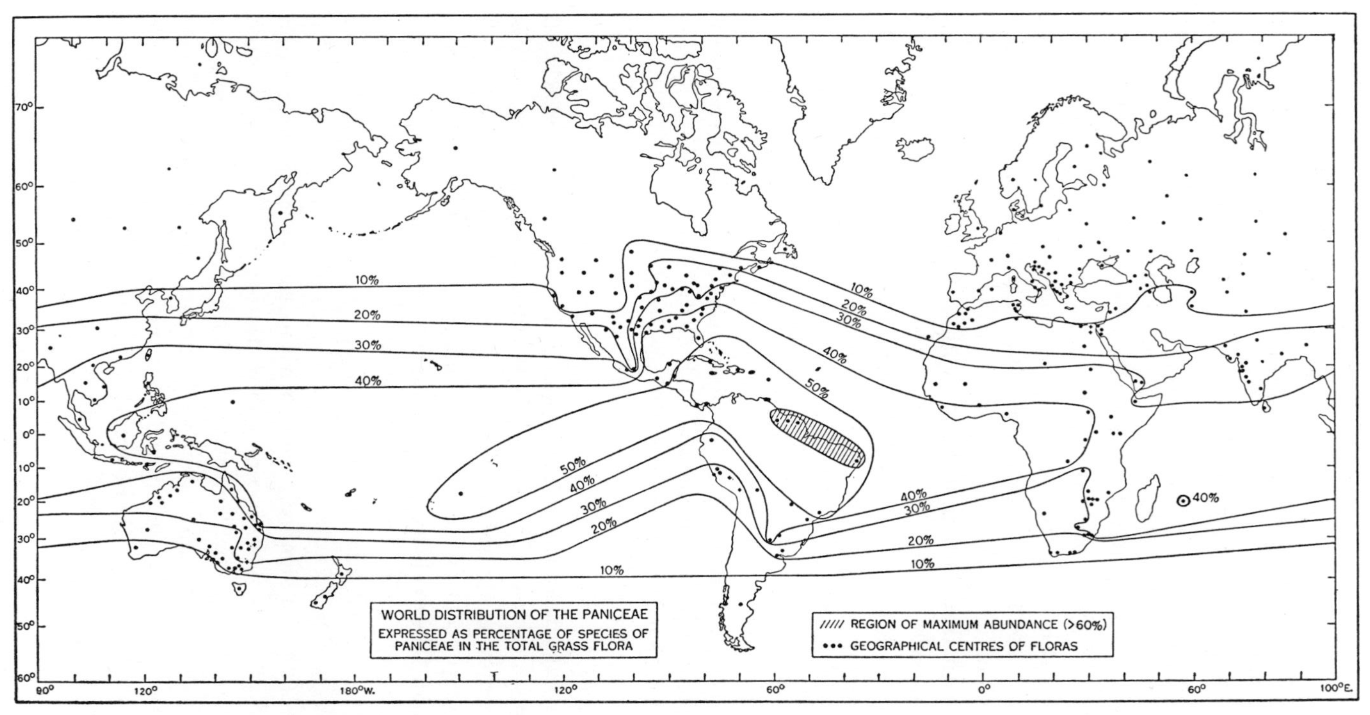

Fig. 26. Map of world distribution of the Paniceae. (From Hartley, 1958*b*.)

tribe lies in southern tropical Africa, and that the large number of species in South America is due to this area being the centre of diversity for certain large genera such as *Paspalum*, *Axonopus* and *Panicum*.

Other tribes show more complex distribution patterns; the Eragrosteae, for instance, are found at super-normal frequency in regions of high winter temperature but are also favoured by a low annual rainfall, i.e. a hot dry climate. They thus possess several disjunct areas of high frequency, primarily in the south and east hemispheres, as in South Africa and parts of Australia.

The legumes have not been studied so thoroughly from this point of view, but there appears to be no such regularity of climatic distribution, although within the sub-family Papilionaceae, which contains nearly all the valuable pasture species, some tribes such as Trifoliae and Viciae are predominantly temperate, and others such as Phaseolae are mainly tropical (Whyte, 1958).

Hartley (1961) has now extended his studies to the distribution of individual genera of grasses. In *Poa*, which contains more than 200 species, the greatest relative specific differentiation occurs in regions of high latitude and high altitude, and there is a close association between a high frequency of *Poa* species and cool summer temperatures. Climatic factors other than temperature appear to have little influence on the distribution. Hartley concludes that the genus has spread successfully to nearly all parts of the world to which it is climatically adapted, and is therefore comparatively ancient; this is borne out by a high degree of complex polyploidy in many of its species (Grun, 1954, 1955). The evidence does not permit firm conclusions about the centre of origin, but there appears to be an ancient centre of species differentiation in the mountain and plateau region of central Asia.

Studies on smaller genera, with more restricted distribution and possibly more recent differentiation, can, however, often indicate a particular centre of diversity. In *Aegilops*, for instance, with comparatively few species, the maximum differentiation occurs in Turkey and northern Iran (Kihara, 1954) while in *Dactylis*, the Mediterranean basin provides the greatest concentration of species or sub-species (Borrill, 1961*a*; Stebbins and Zohary, 1959). Similarly, the eastern Mediterranean is particularly rich in species of *Trifolium* and *Medicago* while northern Iraq, northern Iran and south Turkestan appear to be the main centre of distribution of the perennial species of *Medicago*, *Onobrychis* and *Trigonella*. Again, the American tropics provide the centre of diversity for such valuable legume genera as *Desmodium*, *Phaseolus* and *Centrosema*, as well as for such tropical grasses as *Axonopus*, *Paspalum* and certain sections of *Panicum* (Whyte, 1958).

One may therefore conclude that while in the case of long established taxa, such as tribes or large widespread genera, the present distribution reflects primarily their climatic adaptation, for smaller and more recently developed taxa, a definite centre of diversity, if not of origin, can usually be detected.

We must, however, distinguish between the distribution of wild genera of grasses and legumes, which is determined primarily by climatic and natural biotic factors, and those of cultivated species, which have been deliberately or unconsciously selected by man and his grazing animals. Of the 10,000 or more grass species only about forty are important in sown pastures, and the main centres of distribution of these cultivated grasses fall into three groups (Hartley and Williams, 1956):

(i) Mediterranean, mid-European to central Asian region (twenty-four species including *Dactylis glomerata*, *Lolium perenne*, *Phleum pratense*).
(ii) East tropical Africa (eight species including *Chloris gayana*, *Panicum maximum* and *Pennisetum clandestinum*).
(iii) Sub-tropical South America (four species including *Axonopus affinis*, *Bromus catharticus* and *Paspalum dilatatum*).

Comparatively few cultivated grasses are indigenous to North America.

This distribution evidently reflects not so much the centres of origin of these species, but the regions in which they were taken into cultivation. There is, for instance, a great excess of Festuceae among the cultivated grasses, due to a parallel distribution of this tribe and the early centres of cultivation. Furthermore, most of the valuable sown pasture grasses in both temperate and tropical regions are not natural grassland species, but derive from woodland and forest margin habitats, which have been cleared to form sown grasslands.

The deliberate cultivation of pastures is so recent, that documentary evidence is available of the spread of many of our forage plants. Lucerne (*Medicago sativa*) for instance, is a native of the temperate regions of western Asia, Media and north-west Iran, which have a pronounced continental climate, with a late spring and short hot summer. It reached Greece about 470 B.C. and Italy and North Africa about 150 B.C. and was taken to Spain by the Moors at the beginning of the sixteenth century. The crop then gave rise to two main climatic types; the Mediterranean (southern) group spread into Italy and the south of France and was taken by the Spanish to Mexico, Peru and Chile. The more northern type developed in the Low Countries, northern France and Germany and reached Britain about 1650. Lucerne then reached North America from two sources independently, the cold-susceptible southern types being introduced from Chile and Mexico to the south-western states about 1850, and the winter-hardy Grimm Franconian lucerne from Germany to Minnesota in 1857 (Klinkowski, 1933).

Similarly, red clover (*Trifolium pratense*) which was cultivated in southern Europe in the third and fourth centuries, was re-introduced as a forage crop into Spain in the sixteenth century, and spread via Holland to Germany and England around 1650, developing persistent, winter-hardy populations in the more northern area (Williams, 1945). In the same way Italian rye-grass, derived from the irrigated meadows of Lombardy in the thirteenth to fourteenth centuries, spread to Switzerland in the early 1800's, and was first introduced into Britain in 1831 (Beddows, 1953).

More recently, the settlement of North America, Australia and New Zealand has resulted in a great increase in the distribution of certain forage species, and their adaptation to new environments. In New Zealand, for instance, most of the important forage species, such as perennial ryegrass (*Lolium perenne*), white clover (*Trifolium repens*) and cocksfoot (*Dactylis glomerata*) are of north-west European origin, while in the winter rainfall areas of Australia, it is Mediterranean species such as *Trifolium subterraneum*, *Lolium rigidum* and *Phalaris tuberosa* which are more important. The distribution patterns and local adaptation of many cultivated grasses and legumes are thus extending rapidly at the present time and provide a most valuable illustration of evolution under domestication.

In conclusion, therefore, we can say that because of the relatively recent development of sown pastures, it is possible in the forage grasses and legumes to distinguish between the three consecutive stages of evolution

- (i) the 'natural' distribution of wild grass and legume species in relation to climatic and biotic adaptation,
- (ii) the differentiation of cultivated species under primitive conditions of clearing and cultivation and
- (iii) the present widespread introduction and testing of forage crops throughout the world.

Although the present pattern of distribution of a tribe or genus can often give a useful indication of its past history, this kind of evidence needs to be supplemented by a knowledge of the genetic relationships within the group. Such genetic information can be most usefully obtained from combined cytotaxonomic and hybridization studies.

In the grasses, more than 70 per cent of investigated species are polyploid, and the development of polyploid complexes has played an important part in their evolutionary history. In fact, Stebbins (1956*b*) has used the Gramineae to illustrate the typical development of a polyploid complex. In the simplest case, this originates when a single diploid species extends its range and at the same time differentiates into a number of locally adapted populations or even sub-species. These differentiated diploids hybridize when brought together, and may give rise to tetraploids, which often combine the environmental tolerances of the parent diploids, enabling them to spread into new habitats. These new tetraploids may compete successfully with the parent diploids, which may then decline to a series of local relict populations or even become extinct. The primary polyploids, which often come to behave as functional diploids, can then give rise to a complex series of secondary polyploids by further hybridization.

The above sequence is well illustrated in the forage grasses.

In *Lolium*, for instance, polyploidy has not yet arisen; all species are diploid ($2n = 14$) and the outbreeding species are all fully interfertile (Jenkin, 1954). This diploid group, however, has a wide range in Eurasia, from the Middle East through the Mediterranean to north-west Norway, and has been successfully introduced into most temperate and Mediterranean regions of the world. It contains a wide variety of ecological and agronomic types, ranging from summer-annual to extreme perennial forms, and is one of the most important pasture genera.

Genetic barriers can, however, develop even at the diploid level. The legume, *Trifolium subterraneum*, is widespread in the Mediterranean region and has now become an important pasture plant in Australia. All the material studied is diploid ($2n = 16$ throughout, except for a small local form in Israel with $2n = 12$), but both the Australian and Mediterranean populations fall into a number of incompatible groups between which gene exchange is difficult or even impossible (Morley, 1959). These groups are morphologically similar, and show no regular climatic differentiation. As far as is known, no natural polyploids have arisen from any of these inter-group crosses.

An early stage in the evolution of polyploidy is shown in the genus *Aegilops*. This genus consists of annual weeds of the Mediterranean region and Middle East, and has played an important part in the evolution of the cultivated wheats. Its cyto-taxonomic relationships have therefore been intensively studied (Kihara, 1954). The areas of geographical distribution of the nine diploid species are smaller than those of the tetraploids, and occupy the centre of the area of distribution of the whole genus. Up to four diploid species occur together in parts of Asia Minor and western Iran, and the range of *A. umbellulata*, the diploid which has entered into the constitution of the largest number of tetraploids, overlaps that of most other diploids. In some cases, however, the two ancestral diploids of a particular tetraploid are now widely separated, so that their geographical ranges must have changed considerably since the original hybridization. Since most species are weeds of cereals and overgrown pastures, Kihara suggests that many of the polyploids were formed when early cereal cultivation in Neolithic times brought together different diploids as weeds. The amphiploids appear to be more tolerant of climatic fluctuations than the ancestral diploids, and so would be expected to spread out from the original area. *Aegilops* is evidently a genus in which a polyploid group has developed comparatively recently as a result of the bringing together of the diploids under primitive agriculture, followed by the spread of the polyploids to new ecological niches.

A more complex situation occurs in *Dactylis* (Stebbins and Zohary, 1959; Borrill, 1961*a*). The widely distributed cocksfoot (*Dactylis glomerata*) of Europe and western Asia is a tetraploid ($2n = 28$) and until recently was regarded as an autopolyploid from the only known diploid *D. aschersoniana* ($2n = 14$), found in the forests of north and central Europe. Recent collections from the Near East and the Mediterranean region, however, have revealed a range of different diploid populations, often disjunct in distribution as shown in Fig. 27, and showing quite distinct morphological features. These include such contrasting forms as the tall winter-growing *lusitanica* from the maritime climate of the Portuguese coast, the small stemmy *santai* from Algeria, and the spreading and multinodal *smithii* from the subtropical zone of the Canary Islands, as well as *D. woronowii*, a steppe form from Iran and Turkmenistan and the earlier-known forest type *aschersoniana*, which has also been found in the Himalayas and West China. These diploid populations will all hybridize, but the fertility of the F_1 is usually decreased and the seed weight and seedling vigour also decline. All the tetraploids are highly interfertile (Borrill, 1961*b*). Over most of the range of the genus only tetraploids occur, but in regions where both chromosome forms exist, the tetraploid form often shows considerable resemblance to the local diploid, and in many cases can only be distinguished from it by cytological study.

From such cytotaxonomic evidence, Stebbins (1956*b*) has suggested the following stages in the evolution of *Dactylis*;

(i) the divergence of the diploids in the Tertiary period,
(ii) the bringing together and hybridization of different diploids as a result of climatic changes in the Pleistocene,
(iii) the rapid spread of the tetraploids into new ecological niches, provided both by post-glacial changes, and by forest clearing from Neolithic times onwards.

Many genera of herbage grasses, however, present a more complex picture of secondary polyploidy, as illustrated by the American sections of *Ceratochloa* and *Neobromus* in the genus *Bromus* (Stebbins, 1949). These two sections consist entirely of polyploids and represent relics of a formerly extensive polyploid complex. *Ceratochloa* contains three or four species which

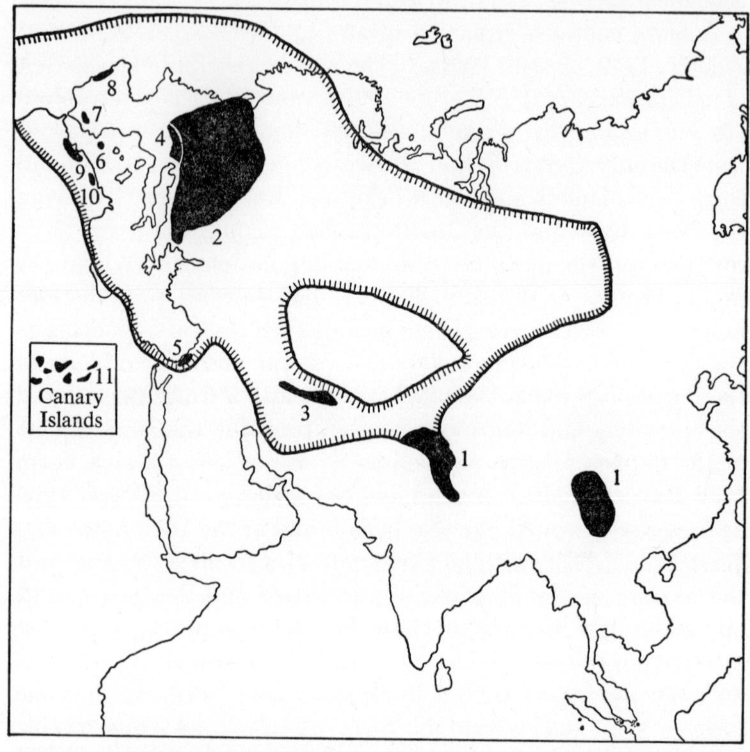

Fig. 27. Generalized distribution of native diploid and tetraploid forms of *Dactylis*: solid, diploid; hatched, boundaries of tetraploid. (1) *D. himalayensis*; (2) *D. aschersoniana*; (3) *D. woronowii*; (4) *D. reichenbachii*; (5) *D. judaica*; (6) *D. ibizensis*; (7) *D. juncinella*; (8) *D. lusitanica*; (9) *D. santai*; (10) *D. mairei*; (11) *D. smithii*.

(From Stebbins and Zohary, 1959.)

are hexaploid ($2n = 42$) and endemic to South America, four octoploids ($2n = 56$) mainly distributed in North America, and one 12-ploid species ($2n = 84$) in south-western U.S.A. *Neobromus*, on the other hand, contains only one species, *B. trinii* which is hexaploid, and native to the arid regions of North and South America. The hexaploid South American species, including *B. catharticus*, are closely related, but behave as functional diploids with no signs of existing diploid or tetraploid ancestors. The octoploids of North America, which include mountain brome (*B. marginatus*) and Californian brome (*B. carinatus*) contain forty-two chromosomes from the South American hexaploids and fourteen from diploids of the Bromopsis section of the genus. They are thus highly successful ancient intersectional hybrids. Finally, *B. arizonicus* ($2n = 84$), is the result of amphiploidy between a hexaploid *B. catharticus* type and the ancient hexaploid *B. trinii* of the section Neobromus. *B. arizonicus* is a common weed over most of its area, and shows that even complex polyploidy can result in adaptation to new ecological niches.

Similar polyploid complexes have been investigated in a number of ancient widespread genera such as *Andropogon* and *Poa*, where the higher polyploids or aneuploids often survive only through apomixis or some form of vivipary (Carnahan and Hill, 1961), and probably remain to be elucidated in *Festuca*, where a range from diploid to decaploid species exists.

Polyploidy does not seem to have played such an important part in the evolution of the forage legumes, in which less than 25 per cent of species are polyploid (Senn, 1938). In *Trifolium*, for instance, which has its highest species concentration in the Mediterranean region, most species are diploid but highly intersterile (Evans, 1962); hybridization between species is very difficult, and genetic barriers may occur even within a single species, as in *T. subterraneum* (Morley, 1959). Such absence of wide hybridization in the legumes may well be associated with the predominant adaptation to insect pollination, as compared to wind pollination in the grasses. Even so, one of the most important temperate forage legumes, *T. repens*, is tetraploid ($2n = 32$), but its ancestral diploids are not known, although Coombe (1962) has recently described a diploid species *T. occidentale* from Cornwall and Devon which is very similar morphologically to *T. repens*.

Cytotaxonomic investigations thus indicate that both primary and secondary polyploidy have played an important part in the evolutionary history of the forage grasses in particular, but that at the same time, considerable evolutionary success has been achieved without polyploidy in certain grasses, such as *Lolium*, and in many legumes, such as *Trifolium*. Such investigations

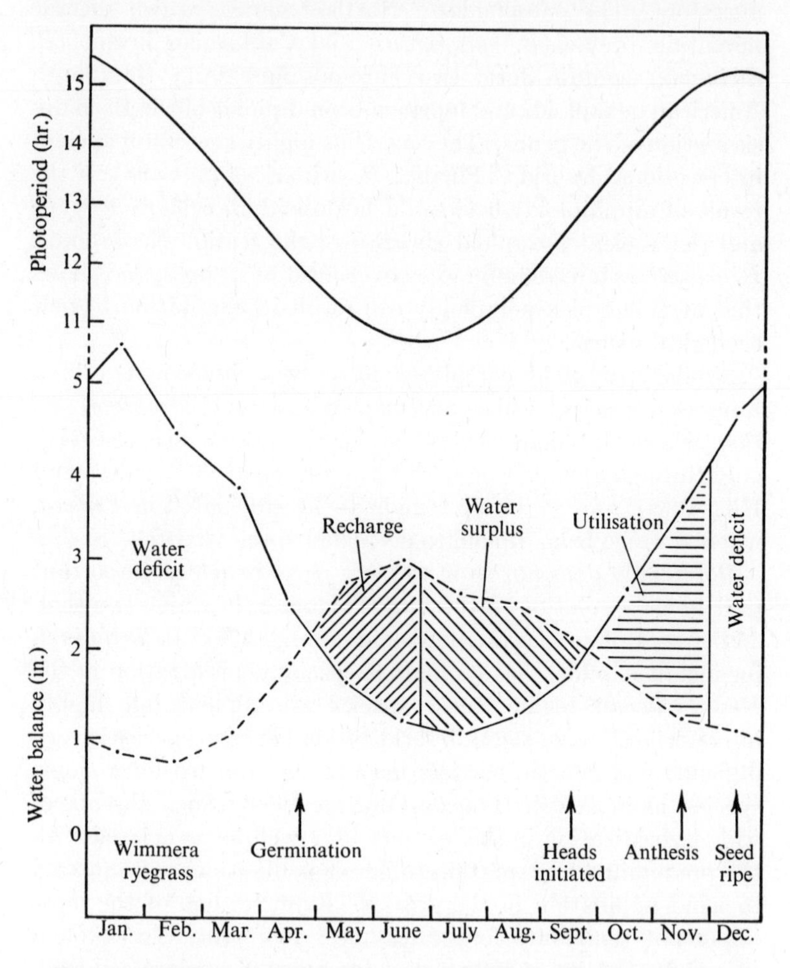

Fig. 28. Relation of growth habit to water supply in Wimmera ryegrass. (From Cooper, 1959*a*.)

while providing valuable information on the past evolution of forage crops, can also indicate to the plant breeder the possibilities of combining desirable genes from two or more related species, or even genera.

Although the pattern of variation between species and genera is based largely on the establishment of genetic barriers, often associated with the development of polyploidy, it is genetic variation *within* the species which is of primary importance for local ecological adaptation and for the purposeful selection of the plant breeder.

Widely ranging forage species usually prove to be built up of a large number of locally adapted populations. The primary adaptation is usually climatic, but the effects of edaphic and agronomic selection are often superimposed on those of local climate. The variation in both seasonal growth rhythm and flowering behaviour shown by many herbage grasses and legumes along a cline from the Mediterranean region to north-western Europe provides a good illustration of such climatic adaptation.

In a Mediterranean environment, the main limiting factor is summer drought, and most species are either winter-annual such as *Lolium rigidum* or *Trifolium subterraneum*, surviving the dry season as seed (as shown in Fig. 28) or, if perennial, show summer dormancy as in *Phalaris tuberosa* or Mediterranean forms of *Dactylis* (Laude, 1953). Winter is the most favourable growing season and most adapted populations possess the ability to grow actively at low temperatures and low light intensities (Cooper, 1962). Recent work, for instance, has shown that many Mediterranean populations of tall fescue, cocksfoot and ryegrass can expand leaf area more rapidly during winter and early spring in Britain than do North European varieties (Borrill, 1961*c*; Chatterjee, 1961) while Barclay (1961) reports that white clover ecotypes from Spain show considerably better winter production in New Zealand than does indigenous material. This active winter production is, however, often associated with susceptibility to frost damage, as shown in Table 9 (Cooper, 1962).

Local varieties from northern Europe, however, show quite a different seasonal pattern of growth. Winter cold is the main limiting factor, and most herbage species are perennials with considerable frost resistance, often associated with winter dormancy. Leaf production at low temperature or low light intensities is poor, but reaches a higher level than in the Mediterranean material during the high light intensities or long photoperiods of the northern summer (Cooper, 1962). Similarly in lucerne, the northern varieties, Canadian, Ladak (from north India) and Spanish Highland, show winter dormancy with consequent high winter survival, while the more southerly forms, including Peruvian, Provence and the Australian Hunter River possess no such winter dormancy and are much more susceptible to frost damage (Morley, Daday and Peak, 1957).

In addition to the seasonal pattern of vegetative growth, the timing of flowering and seed production is usually closely adapted to local climatic conditions. In a Mediterranean environment, for instance, where the length of the growing season is limited by

TABLE 9. *Temperature responses in climatic races of ryegrass and cocksfoot* (from Cooper, 1962)

	Relative increase in leaf area at 5°C	Per cent survival after freezing at −5°C
Ryegrass		
Algiers	26·6	0
New Zealand	13·8	20
Irish	12·2	47
Melle	9·5	57
Pajbjerg	7·6	73
Russian	7·7	92
Cocksfoot		
Bordeaux	29·4	0
Israel	27·9	0
Portugal	24·2	0
Danish	16·4	14
Russian	16·4	33
Norwegian	9·3	88

summer drought, a pattern of flowering responses has usually been selected which results in flowering and seed production at the beginning of the dry season, about the time that the water supply becomes exhausted. In the Mediterranean annuals, *Lolium rigidum* (Cooper 1959*a*, 1960) and *Trifolium subterraneum* (Aitken, 1955*a*, *b*) there is no obligate requirement for cold or short-day exposure before floral induction, and flowering can occur in comparatively short days (10–12 h) as shown in Fig. 28. Local populations differ in their exact flowering requirements, which can often be related to local variations in the length of the possible growing season. Morley (1959) finds a general relationship between the flowering time of subterranean clover populations in Canberra, and the moistness of their original habitat. Similar synchronization of flowering with the beginning of the dry season also occurs in perennial Mediterranean species (Williams, 1956).

In North European material summer drought is not usually important, and the long days of summer provide the optimum conditions for photosynthesis in the ear and for seed development. Floral development in these temperate perennials is usually controlled by a high photoperiodic threshold, and local populations often follow a latitudinal cline in this respect. In red clover, for instance, the Norwegian variety Molstad from 61°N requires very long days or continuous light, while the more southerly Steinacher from Germany (49°N) will flower quite readily in 15 hr. (Schulze, 1957). Similarly, Evans (1939) has shown a close relation between latitude of origin and photoperiod requirements in timothy varieties in the United States. A more complex example is presented by the range grass *Bouteloua curtipendula* in which populations from the southern United States prove to be short-day plants, and those from the northern states to be long-day, with intermediate response from intermediate latitudes (Olmsted, 1944).

In addition to the actual season of flowering, the annual or perennial habit is itself an important adaptive feature. A common basis for the perennial habit in temperate grasses is a winter requirement for either cold or short-day exposure before flowering can occur, thus ensuring a continual succession of vegetative tillers through summer and early autumn. In *Lolium*, for instance, the Mediterranean winter annual *L. rigidum* has no obligatory winter requirement, but most North-European perennial populations must be exposed to some cold or short-day before flowering can take place. This winter requirement is greatest in the extremely persistent types from old grazed pastures (Cooper, 1959*a*, 1960).

Seed dormancy which effectively synchronizes germination with the most favourable climatic season provides another example of climatic adaptation. Most cultivated grasses and legumes are deliberately sown at the season most suitable for germination, and seed dormancy is therefore disadvantageous. In a natural habitat, however, seed production is not necessarily followed by conditions optimum for germination. In a Mediterranean environment, seed dormancy provides insurance against germination at the end of the dry season or during sporadic summer showers. The winter annual *Lolium rigidum*, for instance, shows marked after-ripening dormancy which can be broken by cold, while the cultivated perennial *L. perenne* will germinate immediately after harvesting (Cooper and Ford, unpublished). Similarly, in *Trifolium subterraneum* Morley (1958) finds that germination often requires some exposure to cold and is inhibited by high temperatures. The degree of dormancy of local populations is related to their ecological origins. Dormancy is undesirable where the summer rainfall is well distributed, and unnecessary where the summer rains are absent.

A requirement for cold exposure before germination can occur is also found in some northern European forage grasses from regions with severe winters, as in Arctic populations of *Dactylis glomerata* from North Norway, or hill populations of *Nardus stricta* in Britain (Cooper and Ford, unpublished). In this case, it may effectively prevent autumn germination and hence the winter-killing of very young seedlings.

In cultivated grasses and legumes, however, different farming systems can impose their own selective effects on the primary climatic adaptation, influencing such characters as time of maturity, length of life cycle, and habit of growth. In perennial ryegrass, for instance, the Irish commercial variety has been selected under a system of early harvesting for seed for more than fifty generations, a management which has produced an early-flowering variety with an erect growth habit, and a high proportion of reproductive tillers. Kent perennial ryegrass, on the other hand, is derived from old pastures, which have been hard-grazed in spring and early summer for more than a century; a late-flowering variety with prostrate habit of growth and a high proportion of vegetative tillers has therefore been selected. These differences in flowering habit prove to be based on marked divergence in winter requirement and in photoperiodic response (Cooper, 1959*a*). Similar contrasting local varieties of perennial ryegrass have been developed in recent times in New Zealand under different systems of arable and pasture management (Frankel, 1954).

In the legumes also, the early-flowering broad red clover with a low photoperiodic threshold, has evolved under a system of one year leys, with two or more cuts per year, while the later flowering single cut types have developed under a system of grazing followed by late cutting (Williams, 1945). Similar divergence in relation to management is seen between the cultivated Dutch white clover which has been grown for seed for many years, and the wild white clover indigenous to old grazed pastures.

While the primary adaptation is climatic, with possible agronomic modification, similar population differentiation can occur in response to edaphic factors. Bradshaw and Snaydon (1959) have shown that populations of *Festuca ovina* and of white clover from hill soils of low pH in North Wales are considerably more efficient in growth at low levels of calcium and phosphate than are populations from chalk soils, but that the relative position is reversed at high levels.

A more spectacular demonstration of such edaphic adaptation lies in the development of lead-resistant populations of *Agrostis tenuis* on the waste tips from lead mines in Scotland and North Wales (Bradshaw, 1952; Jowett, 1958). The surrounding pasture populations less than 100 yards away contained no lead-resistant plants, and from a knowledge of the history of these lead mines some quantitative measure of the rate of evolution of the resistant populations could be obtained.

It is evident therefore that in those forage species which have been intensively studied, each local population proves to be highly adapted to its own particular environment. The primary adaptation is usually climatic, but the selective effect of agronomic and edaphic factors is also important. What then is the genetic basis of this close local adaptation, and what are the potentialities for change in such an adapted population, if the direction and intensity of selection are altered?

Most of these adaptive characters which have been investigated genetically, such as seed dormancy in subterranean clover (Morley, 1958) and winter growth in lucerne (Morley, Daday and Peak, 1957) show continuous variation in response to environmental factors and, as might be expected, prove to be polygenically controlled. In the case of flowering behaviour in

Lolium (Cooper, 1954, 1959*b*) and *Trifolium subterraneum* (Davern, Peak and Morley, 1957) both of which have been studied intensively, most of the variation between populations proves to be genetic and additive and there are indications that the individual components, winter inductive requirement and photoperiod response are also under independent polygenic control.

Such polygenic control allows not only the possibility of continuous variation, and hence close local adaptation, but also the production of a similar phenotype by very different combinations of genes (Mather, 1953). Selection in a particular environment is of necessity phenotypic only, and if an adapted population is transferred to a different environment considerable genetic variation between individuals may be revealed. In perennial ryegrass, for instance, the late-flowering pasture variety S.23 heads uniformly outdoors at Aberystwyth (52°N) but diverges considerably in the shorter photoperiods and higher temperatures of the Californian summer, some plants remaining completely vegetative (Peterson, Cooper and Vose, 1958). Similarly, Irish perennial ryegrass, which flowers very uniformly after an autumn sowing, reveals a wide range of heading and non-heading types when sown in early spring, and lines differing greatly in inductive requirement can be selected (Cooper, 1960). The Canadian red clover variety Dollard, which has been selected for uniformity under long days, shows considerable morphological diversity when transferred to short days, individual clones differing greatly in their response to low light intensity (Ludwig, Barrales and Steppler, 1953). In the same way, Vose (1961) has shown that many populations of ryegrass carry hidden genes for resistance to aluminium and manganese toxity, which are revealed only when the plants are grown in conditions of low pH.

Furthermore, most perennial forage grasses and legumes are cross-fertilizing and each individual is therefore highly heterozygous. Considerable variation is thus released in each generation by segregation and recombination and becomes available for selection to work upon. This genetic situation is well illustrated by the results of selection for early and late flowering in Irish and Kent perennial ryegrass. These local populations are quite distinct and differ by about three to four weeks in flowering time. Seven generations of intensive selection have produced lines well outside the original populations and even by the

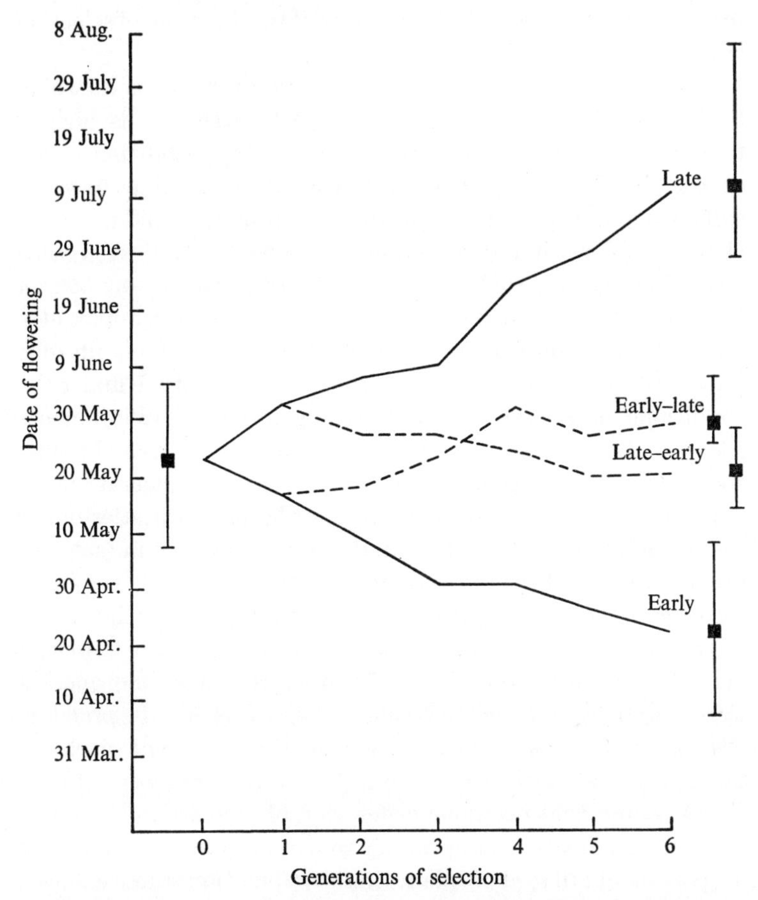

Fig. 29. Response to selection for flowering date in Kent perennial ryegrass. (After Cooper, 1961.)

fourth generation, the complete range of Kent had been obtained from within Irish and vice versa. Response to selection in a Kent stock is shown in Fig. 29 (Cooper, 1961). Similar transgressive response to selection has been obtained for other adaptive characters in ryegrass, including leaf size and rate of leaf production (Cooper and Edwards, 1960) and might be expected in other outbreeding forage species. Evidently, although the population is maintained within its original phenotypic range by stabilizing selection in the original environment, a rapid response can be obtained if the direction of selection is altered.

We thus see that in those cross-fertilizing forage species which have been intensively studied, the genetic structure is such as to maintain close local adaptation through the stabilizing action of selection in each generation, yet at the same time to carry sufficient potential genetic variation to provide rapid adaptive change under altered selection pressures. Even in a self-fertilizing species, such as *Trifolium subterraneum* or *Lolium temulentum*, where most individuals are homozygous, apparently uniform local populations are often found to be built up of a number of different lines, each with similar phenotype under the conditions of selection, but diverging in other environments (Cooper, 1954), and, of course, a single rare outcross in such species can release genetic variation for several generations.

Rapid changes in the performance of herbage varieties under the selective action of climate and management are in fact well known, and stability of type is usually maintained only as long as the direction and intensity of selection during the multiplication of the crop remains constant. Rapid adjustment in flowering behaviour to local conditions of temperature and photoperiod were found by Sylvén (1937) in *Trifolium repens, T. pratense, Phleum pratense* and *Festuca pratensis* in different parts of Sweden, while Smith (1955) working with Ranger lucerne in the U.S.A. found that even one generation of increase in the south resulted in a greater susceptibility to winter injury, and a higher proportion of tall to short plants in the population; these changes were reversed if populations were grown north of their point of origin. Similarly British varieties of white clover and cocksfoot multiplied outside their region of origin are found to differ widely in persistency, habit of growth and time of flowering from the original stocks (Evans, Davies and Charles, 1961).

Even differences in the management of a pasture in one locality can bring about selective change. Charles (1961) has shown that in mixtures of different ryegrass and timothy varieties, the tall erect varieties predominated under lenient defoliation but were severely reduced under frequent grazing. The selection pressure proved to be quite intense in these sown mixtures; only 20 per cent of the initial seedling population was present after two months and only 10 per cent after eleven months. Similarly, Morley (1959) reports that when varietal mixtures of subterranean clover were grown at different locations in Australia, the variety Mount Barker which originally composed only 6 per cent of the population became dominant at most centres. Black (1960) found petiole length to be very important in determining competitive ability for light in this species.

The effects of differential seed production may also be important. Laude and Stanford (1961) found large differences in the relative contributions of different genotypes to the total seed yield of the white clover variety Pilgrim when grown at different locations in the United States or harvested on different dates at the same location.

The most striking examples of genetic change, however, occur when species are introduced into quite new geographical regions, and, as pointed out by Frankel (1954) the introduction of pasture plants into Australia and New Zealand provides a notable opportunity for the study of crop plant evolution in progress. In New Zealand, for instance, perennial ryegrass was established from commercial seed in 1880–90, but by the 1920's two quite distinct populations had evolved from the same original introduction, a long-duration permanent pasture type developed in the North Island under intensive grazing, and a short-lived pseudo-perennial produced in the South Island under arable cropping and short leys. Both these populations differ appreciably from the British commercial variety from which they were derived.

We thus see that many of these outbreeding varieties of forage grasses and legumes can change quite rapidly under altered climate or agronomic selection and potentialities for evolutionary change therefore exist within each species.

Genetic variation, both at and above the species level provides the raw material, not only for evolutionary change but also for the purposeful selection of the plant breeder. How can the breeder best make use of the present pattern of genetic variation in these forage crops?

In the first place, he may be able to base his improvement on locally adapted forage varieties which have already developed under the selective action of appropriate agronomic managements. This approach has been adopted at Aberystwyth in the development of leafy pasture varieties of grasses and legumes based on material from productive grazed pastures, such as S.23 ryegrass derived largely from the fattening pastures of the Midlands and Romney Marsh, and S.123 late-flowering red clover selected from the local Montgomery variety.

Alternatively, where no locally adapted populations are available the breeder can introduce plant material from homologous climatic areas elsewhere. In Australia and New Zealand, for instance, few of the indigenous grasses and legumes are of agronomic value, and nearly all grassland production is based on introduced species. In Australia the most important forage legume, subterranean clover, was first naturalized in Victoria in 1887 and now covers some twenty to thirty million acres, while Wimmera ryegrass (*L. rigidum*) and *Phalaris tuberosa*, also of Mediterranean origin, have proved to be the most valuable forage grasses for winter rainfall areas. The success of these chance introductions has led to the development of an intensive programme of plant collection and introduction from the Mediterranean region. Similarly, in the tropical and sub-tropical areas of Queensland, there is an important place for a suitable legume, and planned introductions have been made from the centres of diversity of such promising genera as *Desmodium* and *Centrosema* (Whyte, 1958).

The value of locally adapted varieties may, however, be restricted by the fact that selection in the past, even under intense grazing, will have been for plant survival, not necessarily for maximum herbage production. Under modern methods of grassland management, the farmer can control fairly closely the time and intensity of cutting or grazing, the nutrient status of the soil, and through irrigation, the water supply, and thus provide the most suitable environment for maximum forage yields. Such an improved agronomic environment may demand the introduction of breeding material from other habitats or clime regions. In Britain, for instance, the use of Mediterranean populations of tall fescue and cocksfoot to provide increased winter production is only possible through the development of appropriate systems of management.

The combination of characters from different climatic populations within the species is a further step in controlled evolutionary change. In lucerne, Morley, Daday and Peak (1957) have been able to combine satisfactorily the high winter production of such southern forms as Hunter River, with the high summer production of the more northerly Canadian and Ladak.

Finally, it may be desirable to combine characters from different species or polyploid races. In *Dactylis*, for instance, detailed cytotaxonomic investigations have made it possible to plan breeding programmes to combine the winter production of the Portuguese diploid *lusitanica* with the winter hardiness of the North European tetraploids (Borrill, 1961*b*). Similar cytotaxonomic information may be required in making use of the winter growing North African tall fescues, some of which prove to be decaploid ($2n = 70$) (Thomas, 1961), while in some high polyploid species of *Poa* Clausen (1952) has been able to stabilize certain valuable interspecific hybrids by the use of apomixis.

A knowledge of the evolutionary history of his crop can thus help the plant breeder at two complementary levels. A cytotaxonomic survey of his material at and above the species level can provide information on the possibilities of hybridization and gene exchange between species or even genera; while a developmental survey of the pattern of genetic variation within the species will often indicate which climatic and agronomic environments can provide the most valuable sources of breeding material. Such information is particularly important in the forage grasses and legumes which are still in active process of domestication.

VIII

CROP PLANT EVOLUTION: A GENERAL DISCUSSION

by J. B. HUTCHINSON

In considering crop plant evolution it is appropriate to begin with the time scale. In the Old World the beginnings of agriculture, which include the beginnings of the culture of the major Old World cereals, the wheats and barleys, can now be identified with some confidence in northern Iraq and neighbouring regions in Iran, Turkey and Jordan. They have been dated at between 7000 and 6000 B.C. Godwin has assembled the evidence for the belief that the practice of agriculture reached north-western Europe about 3000 B.C. Thus the period during which the evolution of our oldest crop plants has gone on, is about 9000 years. In the early part of the period various plants went through the early stage of domestication and were then abandoned—*Chenopodium*, *Polygonum*, *Camelina*—and some late comers were adopted and successfully established—rye and oats. These, having been more recently domesticated, have reached their present state in a shorter time. For the New World the most recent evidence (MacNeish, 1962) indicates that the beginnings of agriculture probably date from about 5000 to 3000 B.C., giving for the New World a maximum evolutionary span of about 7000 years.

Of the crops discussed in these lectures, wheat and barley date from the beginnings of agriculture, and oats and rye from a later period. Doggett has suggested that the domestication of *Sorghum* may have been undertaken by people who brought primitive wheats with them when they migrated into regions where wild *Sorghums* occur, and *Sorghum* must therefore be grouped with the secondary crops. Maize and potatoes are New World crops, and considerable evidence is now available on the beginnings of maize cultivation, giving the earliest date as about 5000 B.C. in Central America. There is no evidence as yet on the antiquity of potato cultivation, but from what is known of the date of the beginnings of agriculture in South America, it seems likely that potatoes have been cropped since 3000 B.C.

It appears, therefore, that the whole vast complex of cultivated plants has evolved in about 9000 years, and some of our advanced and successful crop plants have been developed from their wild ancestors in half that time. The rate of change involved must be greater than in any other group of organisms, and it is not too much to say that for the student of evolution, the crop plants provide the most rewarding material to be found in all biology. There are not only the ancient crops with 9000 years of history, but also the recently adopted crop plants, such as the forage plants where the wild progenitors and the cultivars are still to be found side by side, and even such an evolutionary phenomenon as *Hevea* rubber, of which some individuals of the original domestication are said to be still alive.

Godwin has commented on the wide range of plants that contributed to the diet of early agricultural man, and it is interesting to speculate on the factors that governed the choice of plants for domestication, and determined their success in more advanced agricultural systems. The food needs of mankind may be classified under three heads, carbohydrates, oils and fats, and protein. For his carbohydrate supplies the wide range of plants with which he has experimented has been narrowed down to two main groups, with only a very few plants that do not fall within them. The most important group comprises the grain-producing species of the Gramineae. Second in importance to these is the very diverse group of dicotyledonous roots and tubers. Among the grains, such dicotyledons as *Chenopodium* in the Old World and *Amaranthus* and *Chenopodium* (Quinoa) in the New have given place to the cereals. And such exceptional carbohydrate suppliers as the banana are giving place to roots and tubers, such as yams and sweet potatoes. One might hazard a guess that among grain producers, the highest ratio of grain to vegetative matter is achieved in the monocotyledonous Gramineae, and that an equally favourable ratio of edible to inedible material is only achieved among the dicotyledons among those that produce roots and tubers.

Vegetable oil production comes from monocotyledons and dicotyledons alike, from coconut and oil palm on the one hand, and groundnut, sunflower and the *Brassicas* on the other.

In the provision of the other great human need, protein, the Leguminosae occupy the position held by the Gramineae in the supply of energy. They have a subordinate position in agriculture, partly because the cereals also provide substantial amounts of protein, and partly because they are not as productive as the cereals. Here again, as seed bearers the dicotyledons are not as productive as the monocotyledons.

The choice of wild plants for domestication depended upon their immediate attractiveness. We can see the process going on in the collection and study of grasses and legumes for pasture improvement in Australia, in many developing tropical countries, and even in western Europe and North America. Of those initially selected, only a few are successful and spread, and these only persist so long as they maintain their place in competition with other crop plants. Thus even so long established a forage crop as Sainfoin has almost disappeared, since it did not respond as well as alternative fodder crops to the changing circumstances of British agriculture.

A striking feature of the accounts of crop plant evolution here recorded is the great importance of the continuing genetic contact between the crop plant and its wild relatives. Dodds has described the association between cultivated and wild forms of the potato in South America. Mangelsdorf has drawn attention to the importance of teosinte in the evolution of modern maize. Bell has reported on the persistence of wild relatives of wheat as weeds in wheat fields, and on the opportunities for hybridization and polyploidy that have followed. Doggett has described the situation in the *Sorghum* crop, where weed forms persist in the crop throughout its African range.

In *Sorghum* in particular, the existence and significance of hybridization between the crop plant and its wild relative is apparent to the careful observer, and the effectiveness of gene exchange is evident from the similarity between weed and cultivar throughout the African range of the crop. The maintenance of the distinction between them is an excellent illustration of the strength and effectiveness of 'disruptive' selective forces in a field crop. *Sorghum* appears to have reached India uncontaminated with its weedy relative, and it would be an interesting comparative study to enquire whether the rate of change and the extent of adjustment to agricultural circumstances have been any greater in India without the weed than in Africa in its presence. That change and differentiation are not dependent upon the differentiated gene pool established under disruptive selection is shown by the cottons, in which the crop has evolved in isolation from its wild relatives, and disruptive selection situations are very rare.

One of the major consequences of domestication is an enormous increase in the area of distribution of the species. This is most striking in the spread of crop plants formerly confined to one hemisphere. The spread of wheat and barley to the New World, to South Africa and the high altitude tropical latitudes, and to Australia, and of maize throughout the tropics, subtropics and warm temperate regions of the Old World has all taken place in the last four hundred years. Similar vast extensions in distribution have taken place with most of the important crop plants of the world, and indeed in many cases the chief areas of production are beyond what was the range of the species until quite recent times. Cacao production is concentrated in West Africa, rubber in South-east Asia, bananas in the New World, and cotton in regions of North America well beyond the limits of the species at the time of Colombus.

These vast increases in range of the major crop plants could only have been possible in very plastic organisms. Limits can be discerned beyond which a crop plant cannot be established. Well known examples of such limits to distribution are those to which continental tropical or sub-tropical crops such as maize and soy beans are subject. All attempts to establish them in the agricultural system of the mild, damp, cloudy British Isles have failed, whereas in the drier and hotter summers of continental Europe the selection of well adapted strains has not been difficult. Though limits can be recognized, it is not the limits but the range and variety of the climates in which adaptation is successful that is remarkable. Probably the most remarkable of all is the example of the annual cottons. Six hundred years ago all cottons were perennial shrubs, and since they are all frost susceptible, the crop was entirely confined to frost-free tropical countries. In each of the four cultivated species, however, the range of variability in morphology was such as to permit of the selection of forms that fruited early enough to give a worth-while crop in the first season. These early fruiting cottons were grown in countries with hot summers but cold winters, and the annual habit was imposed on them by frost in the winter. Selection for high productivity completed the process, and now obligate annuals make up almost the whole of the world's crop, and are grown not only in areas with cold winters but also in countries with hot dry seasons which likewise limit the success of the more primitive perennials. Thus in no more than 600 generations, the whole habit of the plant has been changed in response to the limitations of a climate to which the early cultivated forms were quite unsuited.

A similar, though less extreme, change in habit has been established in a much shorter period following the introduction of *Sorghum* into the agriculture of the United States. The common African forms, 8 ft to 12 ft tall, were quite unsuited to mechanized agriculture, and since the produce is a coarse grain used for stock feed, low cost was essential for successful production. Short forms were known, but these also were too tall for convenient mechanical harvesting. In quite a short period, however, *Sorghums* were bred that were short enough for full mechanization, and on them the whole of the present American crop is based. Their genetic structure is well understood, the commercial types carrying a small number of major dwarfing genes, together with a polygene complex that magnifies the major gene effects. It is instructive to note that whereas the major gene effects can be studied conveniently within the American *Sorghum* population, crosses between American dwarfs and Indian agricultural types underwent such complex segregation that the effects of the major genes were lost.

These enormous increases in the distribution of our crop plants have been followed by extremely rapid differentiation of geographical races. The nature and consequences of the differentiation have varied with the circumstances of the particular crops. The nature of the sample of the population from which the original supply of seed was taken has almost always left its mark on the race developed in the new area. Ellis has recently shown that the maize of Nyasaland can be sorted into a group of races that matches fairly closely the races that have been described from north-eastern South America, and the evidence is consistent with the view that the crop is descended from Portuguese introductions from Brazil. Among the cottons of American Upland descent established in the Old World, those of India were derived from the medium staple New Orleans types of the 1860's, and their modern derivatives are shorter in staple than current American types. Those of Africa, on the other hand, came from introductions of a small range of long staple American varieties of the first decade of the present century, and they are still longer in staple than the bulk of the American crop.

The sample that was originally introduced was the raw material of the new race. The race was then fashioned by the selective forces of the new environment, and these have had the most striking effects, in a surprisingly short time. The selection of hairy leaved types in American cottons both in India and in Africa has given rise to a highly distinct phenotype, resistant to the jassid pest that is ubiquitous in the Old World tropics. In cotton, these changes can be fairly accurately timed, and it appears that the development of jassid resistance by natural selection in the fields of the Indian cultivator took no more than about fifty generations. Under selection by plant breeders in southern Africa, the establishment and spread of jassid resistant types was accomplished in a much shorter time. Similar racial differentiation is evident in the potatoes. The European and North American forms have become widely distinct from the types still to be found in the home of the potatoes in South America, even though the number of sexual generations involved must be quite small. Moreover, improvement in South American potatoes now depends heavily on the return to their original home of the modern advanced cultivars developed in the new areas of cultivation.

This complex pattern of spread and differentiation depends upon the existence of a pool of genetic diversity out of which new genotypes emerge under the selection of the diverse environments to which man subjects his crop. In considering the differentiation of the cottons one cannot fail to be impressed by the contrast between the genetic uniformity of the wild species and the enormous genetic diversity of the cultivated species. It seemed necessary to postulate some process by which diversity could arise within a species in the course of domestication and development as a crop plant. Fisher's demonstration that an expanding population would become more variable through the survival of a larger proportion of the naturally occurring mutants provides a statistical basis for increasing variability, and there is good reason to suppose that in other crop plants as well as cotton, the gene pool has been augmented in this way. Nevertheless, it appears that in other crop plants there is not the same disparity between the variability of wild and cultivated species as there is in cotton. Bell has remarked on the variability of the weedy relatives of wheat. Doggett's wild *Sorghums* match the cultivated forms in diversity, though some at least of this may be ascribed to gene exchange. Most important of all, Cooper's forage plants at the very beginning of domestication have within their current limits all the variability that one is accustomed to find in a wide ranging crop plant. Moreover, recent work with *Drosophila* makes it clear that superficial uniformity may conceal an extensive gene pool in a wild species —if indeed *Drosophila* can be regarded as a wild species. Consideration of the time scale discussed above suggests that many stocks of *Drosophila* must have been bred in captivity for as many generations as some of our crop plants, and we ought perhaps to class *Drosophila* with the honey bee in a small category of domestic insects.

The structure of the crop population and the distribution of the variability within it, depends upon the breeding system and the impact of selective forces. Breeding systems in crop plants range from free gene exchange throughout large populations through wind pollination, as in maize, to virtually complete genetic isolation through sterility, with clonal propagation, as in bananas. The wide range of gene exchange through wind pollination is associated with high individual heterozygosity, and the twin phenomena of inbreeding depression and hybrid vigour. A more limited degree of gene exchange occurs in crop plants that are partly self-pollinated and partly cross-pollinated often by insects. Cotton falls in this category, and while the extent of heterozygosity as indicated by segregation on inbreeding, is less than in maize, gene exchange is adequate to ensure the occurrence of the recombination on which progress under selection depends. It serves as a warning against overconfidence in plant breeding techniques to note that the gene exchange from which the modern bacterial blight-resistant Upland cottons of West Africa arose, would not have taken place if the early agricultural botanists had taken the precautions against cross-contamination that we now regard as part of the plant breeder's routine.

Many crop plants, including wheat, barley and oats, are self-pollinated in those parts of their range where the greater part of their crops are produced. They have become virtually mixtures of pure lines, the genetic diversity being partitioned between individuals, and hardly assorted within the genotype of the individuals at all. In the most extreme case, where sexual reproduction has been replaced by clonal propagation, genetic change may be restricted to that which arises by mutation within the clones, and re-assortment of the gene material has then ceased altogether.

In all crop plants the flower structure is such that cross-pollination must have occurred in the past. The Old World cereals have flowers suited to wind pollination. The legumes have the flower pattern characteristic of insect-pollinated plants. Moreover, it is now apparent that at least some of the self-fertilizing crop plants are normally cross-pollinated in their original areas of distribution. The Howards (1909) showed that cross-pollination occurs in wheat in India. Rick (1950) reported cross-pollination in tomatoes in South America and demonstrated the relationship between spread beyond the range of the pollen vectors and the development of a flower type that facilitated self-pollination. Moreover it now appears that the classical Vavilov type of distribution of variability in a crop plant, with a progressive decline in diversity from centre of origin to periphery, is associated with the development of self-pollination. Crops such as wheat and barley lose in variability with every move to a new area, since any pure line not represented in the mother stock for the new crop is lost forever. Cross-pollinated plants such as maize and cotton do not exhibit the Vavilov effect, but on the contrary develop new centres of diversity wherever the crop is established. Indeed it may be that it was this versatility of a cross-pollinating crop that led Anderson to read the Maize story backwards, from Asia to America.

The loss of sexual reproduction must bring evolution in a crop plant as in anything else, almost to a halt. Some increase in clonal diversity will come about by mutation, as is to be seen in the bananas. But with the end of segregation and recombination, there is an end also to the spectacular evolutionary progress characteristic of the crop plants. Only in the banana has the degeneration of the sexual cycle gone to the point where seed production has completely ceased and even in this crop, breeding progress is still possible. The banana breeders have developed an ingenious technique, first proposed by Dodds (1943), whereby they breed superior genomes in seeded, inedible bananas, and then by hybridization add an improved genome as a complete unit to the genome of a sterile, parthenocarpic edible banana. A splendid example of doing good by stealth!

Root and tuber-producing crop plants do not necessarily become seed sterile, and though seed production may be reduced even to vanishing point in commerical crops, it can generally be stimulated sufficiently to make progress by plant-breeding methods possible. There is thus a range in the potential for adaptive response, depending on the breeding system. Outbreeding crop plants may be expected to change in response to changing environments by natural processes alone. Inbreeders will be much less responsive, and adaptation and improvement is only likely through the deliberate intervention of the plant breeder. For those crop plants that are no longer propagated by seed, improvement depends upon the intervention of the plant breeder to re-establish the sexual cycle, though such plants differ from inbreeding seed producers in that the clonal material is generally highly heterozygous, and a seedling progeny consequently segregates widely. We are well aware of the fact that long continued selection to meet the needs of man results in the loss of characters that are essential for survival in the wild. What is perhaps not generally recognized is the further consequence that the breeding system may be so altered—by the development of obligate self-fertilization or by the loss of sexual reproduction—that any further evolutionary changes may also be dependent on the active intervention of man.

The nature of the genetic changes that have gone on during the development of the advanced crop plants is of great importance for an assessment of future prospects. These can be classified under three heads, changes in ploidy, major gene changes, and changes in polygene complexes. Polyploidy is very common among crop plants, but it is evident that increases in chromosome number are by no means necessarily related to crop plant improvement. The most important species of cultivated wheat and oats are polyploid, and polyploidy is common among vegetatively propagated crop plants. Sugar cane, bananas and potatoes are outstanding examples. But development in polyploid wheats has been no greater or more successful than in diploid barley, maize or rice. The tetraploid New World cottons are larger, more vigorous, more productive, and yield produce of higher quality than the diploid species of the Old World. But among the cultivated species of *Phaseolus* the New World species excel those of the Old World in ways that are almost exactly parallel, and the species of both hemispheres are diploid.

It seems probable that the occurrence of polyploidy depends more on the breeding structure of the crop plant than on the selective forces of domestication. In genera of the Gramineae where crosses are possible over a wide range of related species, polyploidy is very common. Moreover, it is common in wild species and in species such as forage grasses that are in the early stages of domestication, as well as in advanced crop plant species such as *Triticum* and *Avena*. In the Leguminosae, on the other hand, crossing between species is difficult, and often impossible, even between those that appear to be closely related, and polyploidy is rare.

The incidence of polyploidy depends not only on the chances of occurrence, but also on the prospects of the polyploid after it has arisen. Riley's demonstration of the genetic nature of the stabilization of meiosis in polyploid wheats is of the greatest importance in elucidating the steps that must occur in the establishment of a successful polyploid. Kimber's (1960) analysis of the data on meiosis in polyploid cottons is sufficient indication that the system of genetic stabilization demonstrated in polyploid wheat may occur in other genera also.

Where vegetative propagation is possible, a polyploid may be successfully established, though its meiotic cycle may be unstable or even very irregular. Thus what has come to be known as 'nobilization' in sugar cane breeding involves exploiting the occurrence of unreduced gametes to increase chromosome number by the addition of a whole set through crossing. The resulting seedling is then propagated solely by cuttings. In such material, progress through plant breeding depends on exploiting the genetic variability in the simpler, sexually normal members of the species or genus, and producing commercially valuable clones by crossing with their complex polyploid relatives. It is then important for the breeder to maintain a clear distinction between his breeding stocks and these commercially acceptable clonal products, the latter being 'dead end' stocks of no further use for breeding.

In considering the importance of major genes in crop plant evolution, it is necessary first to make it clear that all the evidence indicates that genes do not fall into two categories, large and small, but cover the whole range in magnitude of effect. A 'major' gene is no more than a gene having an effect that is large in comparison with the variation due to environmental causes, and hence one that can be identified and studied individually. In this respect, Knight's (1954) analysis of the genetics of blackarm resistance in cotton is important, since in the arid climate with irrigated agriculture in which he worked it was possible to identify a number of genes with effects of different but measurable magnitudes, and to demonstrate the existence of yet others that had too small an effect to be studied individually. Moreover, in the variable climates of the rain-fed regions of Africa, even Knight's major genes became 'minor' genes, in the sense that they could not be individually identified.

There is, nevertheless, a real sense in which changes in major genes have been an instrument of evolutionary change. It is difficult to conceive of the persistence of an unstable polyploid long enough for meiotic stability of the kind demonstrated by Riley to be built up by the selection of a constellation of genes of small individual effect. The major changes which made possible the development of modern maize involved two loci. The origin of spinnable lint from the seed hairs of the wild relatives of the cottons can be accounted for by a change in a single gene. And the difference between brittle and tough rachis in cereals is in general simply inherited. Turning to recent breeding improvement, the short stemmed *Sorghums* bred for combine harvesting in the United States depend basically on a few major dwarfing genes.

It is easy to see that the occurrence of a large mutation would provide an effective stimulus for human selection, whereas the existence of a range of variation in the same character due to minor genes might not. In this way, the establishment of disease resistance, awnlessness in cereals, some forms of pest resistance, besides such quality characters as grain colour and texture, has been brought about in large measure by selection of major genes.

A major change in any important character, however, is rarely adequate in itself. That the American dwarf *Sorghums* have acquired minor genes as well as major dwarfing genes is indicated by the complexity of segregation for height in crosses with tall Indian strains. Major genes for disease resistance seem to be particularly liable to a major responsive change in the parasite, and plant breeders are turning increasingly to minor gene controlled 'field resistance' as more likely to give consistent resistance. In fact the greater part of the genotype of any organism, and of the genes controlling any character, consists of genes of small individual effect. In crosses between related types there may be clear segregation of a single gene, but if the relationship is not close, or if for other reasons there is wide diversity in the segregating populations, the effect of the same major gene may be obscured by the segregation of many other genes having minor effects on the character. To take an example from animal breeding, the transfer of the hornless character from a polled to a normally horned British cattle breed is a simple Mendelian exercise. The establishment of the polled character in a heterogeneous African race of cattle is a much more complex operation, involving minor genes for horn size, differences between the sexes in expression of the character, and changes in dominance relations, all indicative of the segregation of genes with widely different effects. It follows that whatever major changes may arise from simple gene differences, the establishment of a new and superior stock is unlikely to occur unless the minor gene constitution of the material is also subjected to selection, but given the initial improvement conferred by the major gene, further progress of great significance may well follow from selection in the polygenic variation. The efficient manipulation of polygenic variation is only now becoming possible, and in breeding programmes with all kinds of crops, statistical analyses of heritability are likely to lead to much greater exploitation of genes of small individual effect.

Finally, what are the prospects in crop plant evolution? The diversity existing in crop plant species is as great as it has ever been, and with modern travel and the collecting that has gone on since it was first stimulated by Vavilov, the range of material available to the breeder is far greater than ever before. Progress is not limited by variability, but by the selection pressures to which the organism is subject. If these change, the organism changes in response. If they remain the same, an equilibrium is reached, and no further change occurs. Thus in recent years, advances have been made in two main sets of circumstances, in the developing countries of the world where crop plants have been introduced to new environments from elsewhere, and in the western world where there has been a great advance in husbandry practice, leading to a demand—and an opportunity—for the development of new varieties to exploit the new conditions. Perhaps the most striking of recent advances in crop plant performance have been the advances in cereals following the opening up of the North American continent, and the development of an entirely new geographical race following the introduction of one of the species of New World cotton into the Nile valley. On the other hand, it is no accident—and not merely a misguided preoccupation with cash crops—that plant breeders have achieved little in the improvement of African food crops in the past half century. The husbandry of these crops is virtually unchanged and the existing races are thoroughly well adapted to it. That we can breed better *Sorghums*, for instance, when we know what sort of *Sorghum* would be better, is shown by the breeding of new types for new conditions in the United States. Moreover in Africa, when Doggett showed that early cropping, good storing varieties of *Sorghum* would have a place in Sukumaland agriculture in Tanganyika, he had no difficulty in breeding such varieties to his own specification.

The new conditions for which new varieties will be required in the future are likely to arise from improvement in husbandry practices rather than from further changes in crop plant distribution. The spectacular increases in crop production in western countries in recent years have come from a combination of improved husbandry and better varieties. In livestock improvement the saying that 'Half the breed goes in at the mouth' has long been generally accepted. For the crop improver one might suggest as a counterpart 'Half the yield comes up from the soil'. If we accept this situation, and set out to exploit it, we must recognize that breeding improvement will depend first on breeding to a well defined objective, and secondly on acquiring a comprehensive and intimate knowledge of the physiology of the crop to be improved. The well defined breeding objective must be matched with a well conceived blue print. It was because they were easy to define and hence easy to breed for, that disease resistances have figured so largely among the successes of plant breeding. Going a stage further, the response of cereals to high fertility conditions that was easiest to measure was straw stiffness and proneness to lodging. Hence success in breeding for improved husbandry conditions has been largely bound up with improvements in disease resistance and in standing ability. A more detailed physiological approach is now possible, and growth and development studies are being undertaken specifically with the needs of the plant breeder in mind.

These are the conditions that govern the steady improvement of crop plants in response to changing needs and new opportunities. It is interesting to speculate also on the possibility of really major changes in crop plants, with the emergence of new characters. Two that seem both possible and worth while may be discussed briefly. The first is a sheathed *Sorghum* head. *Sorghum* heads do not always emerge completely from the sheath of the topmost leaf. In many parts of Africa losses of grain by bird depredations are very serious, and the fully sheathed maize ear is much better protected than the exposed *Sorghum* head. It might be possible to select a *Sorghum* in which the head never emerges. However, Doggett (personal communication) has pointed out that in the partially sheathed heads that occur already, seed setting is poor, and mildew attack is serious before harvest. Moreover, the sheath would not be large enough to contain a large and fully formed ear. Nevertheless, it seems that all these obstacles must have stood in the way of the development of the maize ear, and they have all been overcome. So even such a radical change is within the limits of reasonable possibility and providing it was recognized as a long term project to be carried on as a minor interest in a more immediate breeding programme, an attempt to assemble the necessary component characters from within the vast diversity of the cultivated *Sorghums* would be worth while.

The second is an enterprise on which a small beginning has already been made, and that is the breeding of a cotton suitable for simple machine harvesting. The major problem in mechanical harvesting of cotton is the separation of 'trash'—debris of leaves, bracts, carpel walls and the like—from the seed cotton. Harvesting would be simple if the seed cotton could be kept away from the parts of the plant that break up into small particles and become entangled in the lint. There are in the diploid species *G. herbaceum*, forms with capsules (bolls) that do not open when ripe. Myers at Stoneville Miss. had *G. hirsutum* (tetraploid) types on to which this boll character had been transferred. Unfortunately their significance was not realized at that time and they were lost, but it is evidently possible to transfer the boll type to the New World cottons. If the closed boll character could be transferred to a commercial Upland cotton, the crop could be left until all bolls were ripe, and indeed until all leaves were killed by frost, and then stripped with a simple machine. After winnowing out all small trash, breaking open the bolls and removing the carpels would be much less difficult, since the broken pieces would be large and hence more easily separated.

In presenting a summary and drawing conclusions after lectures by seven authorities in crop plant evolution, each in his own field, no attempt has been made to add to what they have said. I have tried to pick out some of the general principles that have governed the development and improvement of the world's crop plants, and the temptation then to speculate on what may be done in the future was not to be resisted. Possibly lines other than those suggested will be followed, but there is, beyond doubt, material to hand with which to make changes as great and as significant for human welfare as those that have occurred in the past. Our limitations are the limitations of our scientific insight and imagination, rather than of the biological material with which we work.

BIBLIOGRAPHY

Åberg, E. (1940). The taxonomy and phylogeny of *Hordeum* L. Sect. Cerealia Ands. *Symb. bot. upsaliens*, **4**, (2), 1.

Åberg, E. (1957). Wild and cultivated barleys with pedicelled florets. *Ann. Roy. Agr. Coll. Sweden*, **23**, 315.

Aitken, Y. (1955*a*). Flower initiation in pasture legumes. I. Factors affecting flower initiation in *Trifolium subterraneum* L. *Aust. J. Agric. Res.* **6**, 212.

Aitken, Y. (1955*b*). Flower initiation in pasture legumes. II. Geographical implications of cold temperature requirements of varieties of *Trifolium subterraneum* L. *Aust. J. Agric. Res.* **6**, 245.

Anderson, E. (1947). Field studies of Guatemalan maize. *Ann. Missouri Bot. Gard.* **34**, 433.

Anderson, E. (1949). *Introgressive hybridization.* Wiley and Sons, New York.

Anderson, E. (1953). Introgressive Hybridization. *Biol. Rev.* **28**, 280.

Anderson, E. and Brown, W. L. (1952). The history of the common maize varieties of the United States Corn Belt. *Agric. Hist.* **26**, 2.

Avdulov, N. P. (1931). Karyo-systematische Untersuchungen der Fam. Gramineen. *Bull. appl. Bot. Suppl.* **43**, 259.

Ball, C. R. (1913). The Kaoliangs, a new group of grain sorghums. *U.S.D.A. Bur. Pl. Industr. Bull.* No. **253**.

Barclay, P. C. (1961). Breeding for improved winter pasture production in New Zealand. *Proc. 8th Int. Grassl. Congr.* 326.

Barghoorn, E. S., Wolfe, M. K. and Clisby, K. H. (1954). Fossil maize from the Valley of Mexico. *Bot. Mus. Leafl. Harvard Univ.* **16**, 229.

Bateman, A. J. (1947). The number of *S* alleles in a population. *Nature, Lond.* **160**, 337.

Bear, R. P. (1944). Mutations for waxy and sugary endosperm in inbred lines of dent corn. *J. Amer. Soc. Agron.* **36**, 89.

Beddows, A. R. (1953). The ryegrasses in British agriculture: a survey. *Welsh Pl. Breed. Sta. Bull.* **H17**, 1.

Bell, G. D. H. *et al.* (1955). Investigations in the Triticineae. III. The morphology and field behaviour of the A_2 generation of interspecific and intergeneric amphidiploids. *J. Agric. Sci.* **46**, 199.

Bell, G. D. H. and Lupton, F. G. H. (1962). The breeding of barley varieties. In *The Chemistry and Biology of Barley and Malt*, ed. A. H. Cook, Acad. Press Inc., New York.

Bhatti, A. G. (1959). Cytogenetic studies in *Sorghum vulgare* Pers. and *Sorghum virgatum* (Hack.) Stapf. Hybrids. Ph.D. Thesis, Texas A. & M. Coll. College Station, Texas, U.S.A.

BLACK, J. N. (1960). The significance of petiole lengths, leaf area and light interception in competition between strains of subterranean clover (*Trifolium subterraneum* L.) grown in swards. *Aust. J. Agric. Res.* **11**, 277.

BLEIER, H. (1928). Zytologische Untersuchungen an seltenen Getreide —und Rübenbastarden. *A. F. ind. Abs. u. Vererbgsl., Suppl.* **1**, 447.

BORRILL, M. (1961*a*). The pattern of morphological variation in diploid and tetraploid *Dactylis*. *J. Linn. Soc. (Bot.)*, **56**, 441.

BORRILL, M. (1961*b*). Chromosomal status, gene exchange and evolution in *Dactylis*. I. Gene exchange in diploids and tetraploids. *Genetica*, **32**, 94.

BORRILL, M. (1961*c*). Grass resources for out-of-season production. *Rep. Welsh Pl. Breed. Sta.* 1960, 107.

BOWDEN, W. M. (1959). The taxonomy and nomenclature of the wheats, barleys and ryes and their wild relatives. *Canad. J. Bot.* **37**, 657.

BRADSHAW, A. D. (1952). Populations of *Agrostis tenuis* resistant to lead and zinc poisoning. *Nature, Lond.* **169**, 1098.

BRADSHAW, A. D. and SNAYDON, R. W. (1959). Population differentiation within plant species in response to soil factors. *Nature, Lond.* **183**, 129.

BREGGAR, T. (1928). Waxy endosperm in Argentine maize. *J. Hered.* **19**, 111.

BRIEGER, F. G., GURGEL, J. T. A., PATERNIANI, E., BLUMENSCHEIM, A. and ALLEONI, M. R. (1958). Races of maize in Brazil and other eastern South American countries. *Nat. Acad. Sci., Nat. Res. Counc. Publ.* **593**.

BROOKE, C. (1958). The Durra complex in the Central Highlands of Ethiopia. *Econ. Bot.* **12**, 192.

BROWN, W. L. (1949). Numbers and distribution of chromosome knobs in the United States maize. *Genetics*, **34**, 524.

BRÜCHER, H. (1958). Beiträge zur Abstammung der Kultturkartoffel. *SB dtsch. Akad. Lands. Wis. Berl.* **7**, No. 8, 44.

BUKASOV, S. M. (1939). The origin of potato species. *Physis. B. Aires*, **18**, 41.

BURKILL, I. H. (1937). The races of *Sorghum*. *Bull. Misc. Inf. Kew, Lond.* **112**.

BURKILL, I. H. (1953). Habits of man and the origin of the cultivated plants of the Old World. *Proc. Linn. Soc. Lond.* **164**, 12.

BURTON, G. W. (1951). The adaptability and breeding of suitable grasses for the South Eastern States. *Adv. in Agron.* **3**, 197.

CARDENAS, M. (1956). Origen e historia de la papa. *Tec. agropec., Lima*, **2**, 36.

CARNAHAN, H. L. and HILL, H. D. (1961). Cytology and genetics of forage grasses. *Bot. Rev.* **27**, 1.

CELARIER, R. P. (1958*a*). Cytotaxonomic notes on the subsection Halepense of the genus *Sorghum*. *Bull. Torrey Bot. Cl.* **85**, 49.

CELARIER, R. P. (1958*b*). Cytotaxonomy of the Andropogoneae. III. Sub-tribe Sorgheae, genus *Sorghum*. *Cytologia, Tokyo*, **23**, 395.

Cervantes R., J., Rodríguez, A. and Niederhauser, J. S. (1958). Resistencia al virus causante del achaparramiento del maíz. *Folleto Tecnico*, **29**, 1. Secretaria de Agricultura y Ganaderia, Mexico.

Charles, A. H. (1961). Differential survival of grass cultivars of *Lolium*, *Dactylis* and *Phleum*. *J. Brit. Grassl. Soc.* **16**, 69.

Chatterjee, B. N. (1961). Analysis of ecotypic differences in tall fescue (*Festuca arundinacea* Schreb.). *Ann. Appl. Biol.* **49**, 560.

Clark, J. G. D. (1954). *Excavations at Star Carr*. Cambridge.

Clark, J. G. D. and Godwin, H. (1962). The Neolithic in the Cambridgeshire Fens. *Antiquity*, **36**, 10.

Clausen, J. (1952). New bluegrasses by combining and rearranging genomes of contrasting *Poa* species. *Proc. 6th Int. Grassl. Congr.* **1**, 216.

Clayton, W. D. (1961). (79) Proposal to conserve the generic name *Sorghum* Moench. (Gramineae) versus *Sorghum* Adams. (Gramineae). *Taxon.* **10**(**8**), 242.

Coffman, F. A. (1946). The origin of cultivated oats. *J. Amer. Soc. Agron.* **38**, 983.

Cole, S. (1954). *The Prehistory of E. Africa*. Penguin Books.

Collins, G. N. (1909). A new type of Indian corn from China. *U.S.D.A. Bur. Pl. Ind. Bull.* **161**.

Collins, G. N. (1914). A drought-resisting adaptation in seedlings of Hopi maize. *J. Agric. Res.* **1**, 293.

Collins, G. N. (1920). Waxy maize from Upper Burma. *Science*, **52**, 48.

Collins, G. N. (1921). Teosinte in Mexico. *J. Hered.* **12**, 339.

Coombe, D. E. (1962). *Trifolium occidentale*, a new species related to *T. repens*. *Watsonia*, **5**, 68.

Cooper, J. P. (1954). Studies on growth and development in *Lolium*. IV. Genetic control of heading responses in local populations. *J. Ecol.* **42**, 521.

Cooper, J. P. (1959*a*). Selection and population structure in *Lolium*. I. The initial populations. *Heredity*, **13**, 317.

Cooper, J. P. (1959*b*). Selection and population structure in *Lolium*. II. Genetic control of date of ear emergence. *Heredity*, **13**, 445.

Cooper, J. P. (1960). Short-day and low temperature induction in *Lolium*. *Ann. Bot., Lond.* N.S., **24**, 232.

Cooper, J. P. (1961). Selection and population structure in *Lolium*. V. Continued response and associated changes in fertility and vigour. *Heredity*, **16**, 435.

Cooper, J. P. (1962). Light and temperature responses for vegetative growth. *Rep. Welsh Pl. Breed. Sta.* 1961, 16.

Cooper, J. P. and Edwards, K. J. R. (1960). Selection for leaf area in ryegrass. *Rep. Welsh Pl. Breed. Sta.* 1959, 71.

Correll, D. S. (1948). Collecting wild potatoes in Mexico. *U.S.D.A. Circ.* No. 797, 40.

Correll, D. S. (1962). *The Potato and its Wild Relatives*. Renner: Texas Research Foundation.

Coupland, R. (1938). *East Africa and its Invaders*. Oxford.

CUTLER, H. C. (1944). Medicine men and the preservation of a relict gene in maize. *J. Hered.* **25**, 291.

CUTLER, H. C. and ANDERSON, E. (1941). A preliminary survey of the genus *Tripsacum*. *Ann. Missouri Bot. Gard.* **28**, 249.

DARWIN, C. (1845). *A Naturalist's Voyage*. Journal of researches into the natural history and geology of the countries visited during the voyage of H.M.S. Beagle round the world. London: John Murray.

DAVERN, C. I., PEAK, J. W. and MORLEY, F. H. W. (1957). The inheritance of flowering time in *Trifolium subterraneum* L. *Aust. J. Agric. Res.* **8**, 121.

DOBZANSKY, T. (1955). *Evolution, Genetics, and Man*. New York: Wiley.

DODDS, K. S. (1943). The genetic system of banana varieties in relation to banana breeding. *Emp. J. Exp. Agr.* **11**, 89.

DODDS, K. S. (1962). The classification of cultivated potatoes. In *The Potato and its Wild Relatives* by D. S. Correll. Renner. Texas Research Foundation, p. 517.

DODDS, K. S. and LONG, D. H. (1955). The inheritance of colour in diploid potatoes. *J. Genet.* **53**, 136.

DODDS, K. S. and LONG, D. H. (1956). The inheritance of colour in diploid potatoes. II. The three-factor linkage group. *J. Genet.* **54**, 27.

DODDS, K. S. and PAXMAN, G. J. (1962). The genetic system of cultivated diploid potatoes. *Evolution*, **16**, 154.

ENDRIZZI, J. E. (1957). Cytological studies of some species and hybrids in the *Eu-sorghum*. *Bot. Gaz.* **119**, 1.

ENDRIZZI, J. E. and MORGAN, D. T. Jr. (1955). Chromosomal interchanges and evidence for duplication in haploid *Sorghum vulgare*. *J. Hered.* **46**, 201.

EVANS, A. M. (1962). Species hybridisation in *Trifolium*. I. Methods of overcoming species incompatibility. *Euphytica*, **11**, 164.

EVANS, G., DAVIES, W. E. and CHARLES, A. H. (1961). Shift and the production of authenticated seed of herbage cultivars. *Rep. Welsh Pl. Breed. Sta.* 1960, 99.

EVANS, M. W. (1939). Relation of latitude to certain phases of the growth of timothy. *Amer. J. Bot.* **26**, 212.

FARQUHARSON, L. I. (1957). Hybridization in *Tripsacum* and *Zea*. *J. Hered.* **48**, 295.

FIEDLER, H. and SCHREITER, J. (1959). Das Pachytän-Genom von *Solanum vernei*. *Z. Vererbunglehre*, **90**, 62.

FRANKEL, O. H. (1954). Invasion and evolution of plants in Australia and New Zealand. *Caryologia, Suppl.*, 600.

GALINAT, W. C., MANGELSDORF, P. C. and PIERSON, L. (1956). Estimates of teosinte introgression in archaeological maize. *Bot. Mus. Leafl. Harvard Univ.* **17**, 101.

Galinat, W. C. and Ruppe, R. J. (1961). Further archaeological evidence on the effects of teosinte introgression in the evolution of modern maize. *Bot. Mus. Leafl. Harvard Univ.* **19**, 163.

Garber, E. D. (1950). Cytotaxonomic studies in the genus *Sorghum*. *Univ. Calif. Pub. Bot.* **23**, 283.

Godwin, H. (1944). Age and Origin of The 'Breckland' Heaths of East Anglia. *Nature, Lond.* **154**, 6.

Godwin, H. (1956). *The History of the British Flora.* Cambridge.

Godwin, H. (1960). Prehistoric wooden trackways of the Somerset Levels: their construction, age and relation to climatic change. *Proc. Prehist. Soc.* **26**, 1.

Godwin, H. (1962). Vegetational history of the Kentish Chalk Downs as seen at Wingham and Frogholt. Fertschrift Franz Firbas. *Veröff. Geobot. Inst. Rübel, Zürich,* **37**, 83.

Gottschalk, W. (1954). Die chromosomenstrukter der Solanaceen unter Berücksichtigung phylogenetischer Fragestellungen. *Chromosoma,* **6**, 539.

Gottschalk, W. and Peters, N. (1954). Die chromosomenstrukter als Kriterium für Abstammungsfragen bei Tomate und Kartoffel. *Z. Pflanzenz.* **34**, 71.

Griffiths, D. J. *et al.* (1959). Cytogenetic relationships of certain artificial and natural species of *Avena*. *J. Agric. Sci.* **52**, 189.

Grobman, A., Salhuana, W. and Sevilla, R. *in collaboration with* Mangelsdorf, P. C. (1961). Races of maize in Peru. *Nat. Acad. Sci., Nat. Res. Counc. Publ.* **915**.

Grun, P. (1954). Cytogenetic studies of *Poa*. I. Chromosome numbers and morphology of interspecific hybrids. *Amer. J. Bot.* **41**, 671.

Grun, P. (1955). Cytogenetic studies of *Poa*. II. The pairing of chromosomes in species and interspecific hybrids. *Amer. J. Bot.* **42**, 11.

Hadley, H. H. (1958). Chromosome number, fertility and rhizome expression of hybrids between grain sorghum and Johnson grass (*Sorghum halepense*). *Agron. J.* **50**, 278.

Harborne, J. B. and Corner, J. J. (1961). Plant polyphenols. 4. Hydroxycinnamic acid-sugar derivatives. *Biochem. J.* **81**, 242.

Harlan, J. R. (1961). Geographic origin of plants useful to agriculture, pp. 3–19, *in* Germ Plasm Resources, pp. xii + 381. *Amer. Assoc. Advanc. Sci.*

Hartley, W. (1950). The global distribution of tribes of the Gramineae in relation to historical and environmental factors. *Aust. J. Agric. Res.* **1**, 355.

Hartley, W. (1958*a*). Studies on the origin, evolution and distribution of the Gramineae. I. The Tribe Andropogoneae. *Aust. J. Bot.* **6**, 116.

Hartley, W. (1958*b*). Studies on the origin, evolution and distribution of the Gramineae. II. The Paniceae. *Aust. J. Bot.* **6**, 343.

Hartley, W. (1961). Studies on the origin, evolution, and distribution of the Gramineae. IV. The genus *Poa*. *Aust. J. Bot.* **9**, 152.

HARTLEY, W. and WILLIAMS, R. J. (1956). Centres of distribution of cultivated pasture grasses and their significance for plant introduction. *Proc. 7th Int. Grassl. Congr.* 190.

HAWKES, J. G. (1956*a*). Taxonomic studies on the tuber-bearing Solanums. I. *Solanum tuberosum* and the tetraploid species complex. *Proc. Linn. Soc. Lond.* **166**, 97.

HAWKES, J. G. (1956*b*). A revision of the tuber-bearing Solanums. *Ann. Rep. Scot. Pl. Breed. Sta.* 37.

HAWKES, J. G. (1958). Kartoffel. I. Taxonomy, cytology and crossability. *Handbuch der pflanzenzüchtung*, **3**, 1.

HECTOR, J. M. (1936). *Introduction to the Botany of Field Crops*, Vol. I. Cereals. Central News Agency, Ltd., Johann.

HELBAEK, H. (1950). Tollund mandens sidste maaltid. *Aarbøger for Nordisk Oldkyndighed og Historie.*

HELBAEK, H. (1953). Early crops in Southern England. *Proc. Prehist. Soc.* **18**, 194.

HELBAEK, H. (1954). Prehistoric Food Plants and Weeds in Denmark. *Danm. Geologiske Undersøgelse*, II R, 80.

HELBAEK, H. (1955). Ancient Egyptian wheats. *Proc Prehist. Soc.* **21** (N.S.), 93.

HELBAEK, H. (1958). Grauballemandens sidste Maaltid. *Kuml. Årbog for Jysk Arkaeologisk Selskab.*

HELBAEK, H. (1959). Domestication of food plants in the Old World. *Science*, **130**, 365.

HELBAEK, H. (1960*a*). Comment on Chenopodium as a food plant in Prehistory. *Ber. d. Geobot. Inst. Rübel, Zürich*, **31**, 16.

HELBAEK, H. (1960*b*). Paleoethnobotany of the Near East and Europe, in *Prehistoric Investigations in Iraqi Kurdistan*, by R. J. Braidwood and B. Howe (No. 31 in series Studies in Ancient Oriental Civilization), pp. 99–118. Univ. of Chicago Press.

HOROWITZ, S. and MARCHIONI, A. H. (1940). Herencia de la resistencia a la langosta en el maiz. 'amargo'. *An. Inst. Fito. Santa Catalina*, **2**, 27.

HOUGAS, R. W., PELOQUIN, S. J. and ROSS, R. W. (1958). Haploids of the common potato. *J. Hered.* **49**, 103.

HOWARD, A. and HOWARD, G. L. C. (1909). The varietal characteristics of Indian wheats. *Mem. Dept. Agr. Ind.* (Bot. Ser.), II, No. 7.

HOWARD, H. W. (1960). Potato cytology and genetics, 1952–1959. *Bibliogr. Genet.* **19**, 87.

HUSKINS, C. L. (1931). A cytological study of Vilmorin's unfixable dwarf wheat. *J. Genet.* **25**, 113.

IVERSEN, J. (1941). Landnam i Danmarks Stenalder. *Danm. Geol. Unders.* R. II, Nr. 66.

JAIN, S. K. (1960). The genetics of rye (*Secale* spp.). *Bibliogr. Genet.* **19**, 1.

JENKIN, T. J. (1954). Interspecific and intergeneric hybrids in herbage grasses. VII. *Lolium perenne* L. with other *Lolium* species. *J. Genet.* **52**, 300.

JESSEN, K. and HELBAEK, H. (1944). Cereals in Great Britain and Ireland in prehistoric and early historic time. *K. Danske Vidensk. Selsk. Biol. Skr.* **3**, 2.

JONES, E. T. (1956). The origin, breeding and selection of oats. *Agric. Rev.* **2** (1), 20.

JOWETT, D. (1958). Populations of *Agrostis* spp. tolerant of heavy metals. *Nature, Lond.* **182**, 816.

JUZEPCZUK, S. W. and BUKASOV, S. M. (1929). (A contribution to the question of the origin of the potato.) *Proc. U.S.S.R. Congr. Genet. Pl. and An. Breed.* **3**, 593.

KAMM, A. (1954). The discovery of wild six-row barley and wild *Hordeum intermedium* in Israel. *Ann. Roy. Agric. Coll., Sweden*, **21**, 287.

KEMPTON, J. H. (1936). Maize as a measure of Indian skill. *Univ. N. Mex. Bull.* **296**, 19.

KEMPTON, J. H. and POPENOE, W. (1937). Teosinte in Guatemala. *Carnegie Inst. Washington Publ.* **483**, 199.

KIDD, H. J. (1958). Quoted in ANDERSON, E. Genetics in plant breeding. *Brookhaven Symposia in Biology*, No. 9, 123.

KIESSELBACH, T. A. and KEIM, F. D. (1921). The regional adaptation of corn in Nebraska. *Nebraska Agric. Exp. Sta. Res. Bull.* **19**, 1.

KIHARA, H. (1924). Cytologische und genetische Studien bei wichtigen Getreidearten mit besonderer Rucksicht auf das Verhalten der Chromosomen und die Sterilität in den Bastarden. *Mem. Coll. Sci., Kyoto Imp. Univ.* **B**, 1.

KIHARA, H. (1937). Genomanalyse bei *Triticum* und *Aegilops*. VII. Kurze Uebersicht über die Eryebiusse der Jahre, 1934–36. *Mem. Coll. Agr., Kyoto Imp. Univ.* **41**, 1.

KIHARA, H. (1954). Considerations on the evolution and distribution of *Aegilops* species based on the analyser method. *Cytologia*, **19**, 336.

KIHARA, H. and LILIENFELD, F. (1949). A new synthesized 6*x*-wheat. *Proc. Xth Int. Congr. Genet.* 307.

KIHARA, H. and TANAKA, M. (1958). Morphological and physiological variation among *Aegilops squarrosa* strains collected in Pakistan, Afghanistan and Iran. *Preslia*, **30**, 241.

KIMBER, G. (1960). The association of chromosomes in haploid cotton. *Heredity*, **15**, 453.

KLINKOWSKI, M. (1933). Lucerne: its ecological position and distribution in the world. *Imp. Bur. Pl. Genet.* (*Herb. Pl.*) Bull. **12**, 1.

KNIGHT, R. L. (1954). Cotton breeding in the Sudan. I. Egyptian cottons. *Emp. J. Exp. Agr.* **22**, 68.

KULESHOV, N. N. (1928). Some peculiarities in the maize of Asia. *Bull. Appl. Bot. and Pl. Breed.* **19**, 325.

KURZ, E. B., LIVERMAN, J. L. and TUCKER, H. (1960). Some problems concerning fossil and modern corn pollen. *Bull. Torrey Bot. Cl.* **87**, 85.

LAUDE, H. M. (1953). The nature of summer dormancy in perennial grasses. *Bot. Gaz.* **114**, 284.

LAUDE, H. M. and STANFORD, E. H. (1961). Environmentally induced changes in gene frequency in a synthetic forage variety grown outside the region of adaptation. *Proc. 8th Int. Grassl. Congr.* 180.

LEAKEY, L. S. B. (1931). *The Stone Age Cultures of Kenya Colony.* Cambridge University Press.

LEAKEY, K. O. and LEAKEY, L. S. B. (1950). *Excavations at the Njoro River Cave.* Oxford University Press.

LILIENFELD, F. A. (1951). H. Kihara: Genome-Analysis in *Triticum* and *Aegilops*. X. Concluding review. *Cytologia,* **16**, 101.

LINNEUS, C. (1953). *Species Plantarum.* Stockholm.

LUDWIG, R. A., BARRALES, H. G. and STEPPLER, H. (1953). Studies on the effect of light on the growth and development of red clover. *Canad. J. Agric. Sci.* **33**, 274.

MCFADDEN, E. S. and SEARS, E. R. (1946). The origin of *Triticum spelta* and its free-threshing hexaploid relatives. *J. Hered.* **37**, 81, 107.

MCKEE, R. (1962). *R*-genes in *Solanum stoloniferum. Euphytica,* **11**, 42.

MACKEY, J. (1954*a*). Neutron and X-ray experiments in wheat and a revision of the speltoid problem. *Hereditas,* **40**, 65.

MACKEY, J. (1954*b*). The taxonomy of hexaploid wheat. *Svensk. bot. Tidskr.* **48**, 579.

MACKEY, J. (1958). Mutagenic response in *Triticum* at different levels of ploidy. *Proc. First Int. Wheat Genet. Symp.* 88.

MACNEISH, R. (1961). Restos precerámicos de la Cueva de Coxcatlán en el sur de Puebla. *Inst. Nac. Anthrop. Hist.* **10**, 5.

MACNEISH, R. S. (1962). 2nd Ann. Rep. of the Tehuacan Archaeological-Botanical Project. R. S. Peabody Foundation, Andover, Mass.

MAGUIRE, M. P. (1957). Cytogenetic studies of *Zea* hyperploid for a chromosome derived from *Tripsacum. Genetics,* **42**, 473.

MALHEIRAS-GARDÉ, N. (1959). Mechanisms of species isolation in tuberous *Solanum. Agron. lusit.* **21**, 19.

MALZEW, A. I. (1930). The genus Eu-avena. *Bull. app. Bot. Pl. Breed. Leningr.* Suppl. No. 38.

MANGELSDORF, P. C. (1924). Waxy endosperm in New England maize. *Science,* **60**, 222.

MANGELSDORF, P. C. (1950). The mystery of corn. *Sci. Amer.* **183**, 20.

MANGELSDORF, P. C. (1951). Hybrid corn: its genetic basis and its significance in human affairs. In *Genetics in the 20th Century,* p. 555. New York: Macmillan.

MANGELSDORF, P. C. (1952). Hybridization in the evolution of maize. In *Heterosis,* p. 175. Iowa State College Press.

MANGELSDORF, P. C. (1958). The mutagenic effect of hybridizing maize and teosinte. *Cold Spring Harbor. Symp. Quant. Biol.* **23**, 409.

MANGELSDORF, P. C. (1961). Introgression in maize. *Euphytica,* **10**, 157.

MANGELSDORF, P. C. and CAMERON, J. W. (1942). Western Guatemala, a secondary center of origin of cultivated maize varieties. *Bot. Mus. Leafl. Harvard Univ.* **10**, 217.

MANGELSDORF, P. C. and LISTER, R. H. (1956). Archaeological evidence on the evolution of maize in northwestern Mexico. *Bot. Mus. Leafl. Harvard Univ.* **17**, 151.

MANGELSDORF, P. C., MACNEISH, R. S. and GALINAT, W. C. (1956). Archaeological evidence on the diffusion and evolution of maize in northeastern Mexico. *Bot. Mus. Leafl. Harvard Univ.* **17**, 125.

MANGELSDORF, P. C. and MANGELSDORF, HELEN P. (1957). Genotypes involving the *Tu-tu* locus compared in isogenic stocks. *Maize Gen. Coöp News Letter*, **31**, 65.

MANGELSDORF, P. C. and OLIVER, D. L. (1951). Whence came maize to Asia? *Bot. Mus. Leafl. Harvard Univ.* **14**, 263.

MANGELSDORF, P. C. and REEVES, R. G. (1939). The origin of Indian corn and its relatives. *Texas Agric. Exp. Sta. Bull.* **574**.

MANGELSDORF, P. C. and REEVES, R. G. (1959*a*). The origin of corn. I. Pod corn the ancestral form. *Bot. Mus. Leafl. Harvard Univ.* **18**, 329.

MANGELSDORF, P. C. and REEVES, R. G. (1959*b*). The origin of corn. IV. Place and time of origin. *Bot. Mus. Leafl. Harvard Univ.* **18**, 413.

MANGELSDORF, P. C. and SMITH, C. E. Jr. (1949). New archaeological evidence on evolution in maize. *Bot. Mus. Leafl. Harvard Univ.* **13**, 213.

MARKS, G. E. (1955). Cytogenetic studies in tuberous *Solanum* species. I. Genomic differentiation in the group Demissa. *J. Genet.* **53**, 262.

MARKS, G. E. (1958). Cytogenetic studies in tuberous *Solanum* species. II. A synthesis of *Solanum* × *vallis-mexici* Juz. *New Phytol.* **57**, 300.

MATHER, K. (1953). The genetical structure of populations. *Symp. Soc. Exp. Biol.* **7**, 66.

MAUNY, R. (1953). Principales Plantes Cultives d'Afrique Occidentale. *Bull. Inst. Franc. Afr. Noire*, **15**, 684.

MORLEY, F. H. W. (1958). The inheritance and ecological significance of seed dormancy in subterranean clover (*Trifolium subterraneum* L.). *Aust. J. Biol. Sci.* **11**, 261.

MORLEY, F. H. W. (1959). Natural selection and variation in plants. *Cold Spr. Harb. Symp. Quant. Biol.* **24**, 47.

MORLEY, F. H. W., DADAY, H. and PEAK, J. W. (1957). Quantitative inheritance in lucerne (*Medicago sativa* L.). I. Inheritance and selection for winter yield. *Aust. J. Agric. Res.* **8**, 655.

MURDOCK, G. P. (1959). *Africa, its Peoples, and their Culture History.* McGraw-Hill, New York.

OCHOA, C. (1958). Expedicion colectora de papas cultivadas a la cuenca del Lago Titicaca. I. Determinacion sistematica y numero chromosomico del material colectado. PCEA Ministerio de Agricultura, Lima, Peru.

OINUMA, T. (1952). Karyomorphology of cereals. *Biol. J. Okayama Univ.* **1**, 12.

OKAMOTO, M. (1962). Identification of the chromosomes of common wheat belonging to the *A* and *B* genomes. *Can. J. Genet. Cytol.* **4**, 31.

OLMSTED, C. E. (1944). Growth and development in range grasses. IV. Photoperiodic responses in twelve geographic strains of side-oats grama. *Bot. Gaz.* **106**, 46.

PANDEY, K. K. (1960). Self-incompatibility system in two Mexican species of *Solanum*. *Nature, Lond.* **185**, 483.

PELOQUIN, S. J. and HOUGAS, R. W. (1960). Genetic variations among haploids of the common potato. *Amer. Potato J.* **37**, 289.

PERLOVA, R. L. (1958). (The behaviour of wild and cultivated species of potato in different geographical regions of the Soviet Union.) *Akad. Nank SSSR. Glarn. Bot. Sad. Moscow*, 1958, 238 (Russian).

PETERSON, M. L., COOPER, J. P. and VOSE, P. B. (1958). Non-flowering strains of herbage grasses. *Nature, Lond.* **181**, 591.

PIEDALLU, A. (1923). *Le Sorgho, son Histoire, ses Applications*. Paris, Challamel.

RAJHATHY, T. and MORRISON, J. W. (1960). Genome homology in the genus *Avena*. *Canad. J. Genet. Cytol.* **2**, 278.

RANDOLPH, L. F. (1955). History and origin of corn. II. Cytogenetic aspects of the origin and evolutionary history of corn. In *Corn and Corn Improvement*, p. 16. New York: Academic Press.

RAPER, J. R., KRONGELB, G. S. and BAXTER, M. G. (1958). The number and distribution of incompatibility factors in *Schizophyllum*. *Amer. Nat.* **92**, 221.

RICK, C. M. (1950). Pollination relations of *Lycopersicon esculentum* in native and foreign regions. *Evolution*, **4**, 110.

RILEY, R. (1960). The diploidisation of polyploid wheat. *Heredity*, **15**, 407.

RILEY, R. and BELL, G. D. H. (1958). The evaluation of synthetic species. *Proc. First Int. Wheat Genet. Symp.* 161.

RILEY, R. and CHAPMAN, V. (1958*a*). The production and phenotypes of wheat-rye chromosome addition lines. *Heredity*, **12**, 301.

RILEY, R. and CHAPMAN, V. (1958*b*). Genetic control of the cytologically diploid behaviour of hexaploid wheat. *Nature, Lond.* **182**, 713.

RILEY, R. and CHAPMAN, V. (1960). The *D* genome of hexaploid wheat. *Wheat Inf. Service*, **11**, 18.

RILEY, R., KIMBER, G. and CHAPMAN, V. (1961). The origin of the genetic control of the diploid-like behaviour of polyploid wheat. *J. Hered.* **52**, 22.

RILEY, R., UNRAU, J. and CHAPMAN, V. (1958). Evidence of the origin of the *B* genome of wheat. *J. Hered.* **49**, 91.

ROBERTS, L. M., GRANT, U. J., RAMIREZ E., R., HATHEWAY, W. H. and SMITH, D. L. *in collaboration with* MANGELSDORF, P. C. (1957). Races of maize of Colombia. *Nat. Acad. Sci., Nat. Res. Counc. Publ.* **510**.

Sachs, L. (1953). Chromosome behaviour in species hybrids of *Triticum timopheevi*. *Heredity*, **2**, 49.

Saint-Hilaire, A. de (1829). Lettre sur une variété remarquable de maís du Bresil. *Ann. Sci. Nat.* **16**, 143.

Sakamura, T. (1918). Kurze Mitteilung über die Chromosomenzahlen und die Verwandtschaftsverhaltnisse der *Triticum*—Arten. *Bot. Mag., Tokyo*, **32**.

Salaman, R. N. (1949). *The History and Social Influence of the Potato*. Cambridge University Press.

Salaman, R. N. (1954). The origin of the early European potato. *J. Linn. Soc.* (*Bot.*), **55**, 185.

Salaman, R. N. and Hawkes, J. G. (1949). The character of the early European potato. *Proc. Linn. Soc. Lond.* **161**, 71.

Sarkar, P. and Stebbins, G. L. (1956). Morphological evidence concerning the origin of the *B* genome in wheat. *Amer. J. Bot.* **43**, 297.

Sax, K. and Sax, H. J. (1924). Chromosome behaviour in a genus cross. *Genetics*, **9**, 454.

Schick, R., Schick, E. and Haussdörfer, M. (1958). Ein beitrag zur physiologischen Spezialisierung von *Phytophthora infestans*. *Phytopath. Z.* **31**, 225.

Schiemann, E. (1948). *Weizen, Roggen, Gerste*. Fischer, Jena.

Schreiber, K. (1954). Die Glykoalkaloide de Solanaceen. *Chem. Tech.* **12**, 648.

Schulze, E. (1957). (Photoperiodic experiments on perennial fodder plants I.). Photoperiodische Versuche an mehrjährigen Futterpflanzen. I. Mitteilung. *Z. Acker- u. PflBau.* **103**, 198.

Sears, E. R. (1948). The cytology and genetics of the wheats and their relatives. *Advanc. Genet.* **2**, 239.

Sears, E. R. (1954). The aneuploids of common wheat. *Res. Bull. Mo. Agr. Exp. Sta.* No. 572.

Sears, E. R. (1956). The *B* genome of *Triticum*. *Wheat Inf. Service*, **4**, 8.

Sears, E. G. (1958). The aneuploids of common wheat. *Proc. First Int. Wheat Genet. Symp.* 221.

Sears, E. G. (1959). The systematics, cytology and genetics of Wheat. *Handbuch der Pflanzenzuchtung*, **2**, 164.

Seddon, B. (1960). Excavations at Dinas Emrys, Beddgelert (Caern.): report on the organic deposits in the port at Dinas Emrys. *Archaeologia Cambrensis*, 13.

Senn, H. A. (1938). Chromosome number relationships in the Leguminoseae. *Bibl. Genet.* **12**, 175.

Shebeski, L. H. (1958). Speculations on the impact of the *D* genome. *Proc. First Int. Wheat Genet. Symp.*, 237.

Smith, D. (1955). Influence of area of seed production on the performance of Ranger alfalfa. *Agron. J.* **47**, 201.

Smith, L. (1951). Cytology and genetics of barley. *Bot. Rev.* **17**, 1, 133, 285.

Snowden, J. D. (1936). *The Cultivated Races of Sorghum*. Adlard and Son, London.

SNOWDEN, J. D. (1955). The wild fodder Sorghums of the genus *Eu-Sorghum*. *J. Linn. Soc. Lond.* **55**, 191.

SPRAGUE, G. F. (1939). An estimation of the number of top-crossed plants required for adequate representation of a corn variety. *J. Amer. Soc. Agron.* **31**, 11.

STAPF, O. (1917). In PRAIN, D. *Flora of Tropical Africa*, vol. 9. Reeve & Co., London.

STEBBINS, G. L. (1949). Evolutionary significance of natural and artificial polyploids in the family Gramineae. Proc. 8th Int. Congr. Genet. Stockholm 1948. *Hereditas* (suppl.), 461.

STEBBINS, G. L. (1956*a*). Taxonomy and the evolution of genera with special reference to the family Gramineae. *Evolution*, **10**, 235.

STEBBINS, G. L. (1956*b*). Cytogenetics and evolution of the grass family. *Amer. J. Bot.* **45**, 890.

STEBBINS, G. L. and ZOHARY, D. (1959). Cytogenetic and evolutionary studies in the genus *Dactylis*. I. Morphology, distribution and interrelationships of the diploid subspecies. *Univ. Calif. Publ. Bot.* **31**, 1.

STONOR, C. R. and ANDERSON, E. (1949). Maize among the hill peoples of Assam. *Ann. Missouri Bot. Gard.* **36**, 355.

STURTEVANT, E. L. (1894). Notes on Maize. *Bull. Torrey Bot. Cl.* **21**, 319 and 503.

SWAMINATHAN, M. S. and HOWARD, H. W. (1953). The cytology and genetics of the potato (*Solanum tuberosum*) and related species. *Bibl. Genet.* **16**, 1.

SWAMINATHAN, M. S. and RAO, M. V. P. (1961). Macro-mutations and sub-specific differentiation in *Triticum*. *Wheat Inf. Service*, **13**, 9.

SYLVÉN, N. (1937). The influence of climatic conditions on type composition. *Imp. Bur. Pl. Genet.* (*Herb. Pl.*) Bull. **21**, 1.

TACKHOLM, V., TACKHOLM, G. and DRAR, M. (1941). *Flora of Egypt*. I, 538. Cairo.

TAKAHASHI, R. (1955). The origin and evolution of cultivated barley. *Advanc. Genet.* **7**, 227.

THODAY, J. M. and BOAM, T. B. (1961). Effects of disruptive selection. V. Quasi-Random mating. *Heredity*, **16**, 219.

THOMAS, P. T. (1961). Chromosomes and grass breeding. *J. Brit. Grassl. Soc.* **16**, 276.

TURNER, J. (1962). The Tilia decline: an anthropogenic interpretation. *New Phytol.* **61**, 328.

VOSE, P. B. (1961). Intraspecific differences in nutrient uptake and assimilation. *Rep. Welsh Pl. Breed. Sta.* 1960, 22.

WAGENAAR, E. B. (1961). Studies on the genome constitution of *Triticum timopheevi* Zhuk. I. Evidence for genetic control of meiotic irregularities in tetraploid hybrids. *Can. J. Genet. Cytol.* **3**, 47.

WAGENHEIM, K-H. F. v., FRANSDEN, N. O. and ROSS, H. (1957). Über neue Ergebnisse zur Cytologie und verwandte Fragen bei *Solanum*. *Z. Pflanzenz.* **37**, 41.

Walker, D. (1955). Skelsmergh Tarn and Kentmere, Westmorland. *New Phytol.* **54**, 222.

Wallace, H. A. and Brown, W. L. (1956). *Corn and its Early Fathers.* Michigan State University Press.

Watkins, A. E. (1930). The wheat species: a critique. *J. Genet.* **23**, 173.

Watt, G. (1893). *Dictionary of the Economic Products of India,* **6**, 291. London and Calcutta.

Weatherwax, P. (1942). The Indian as a corn breeder. *Proc. Indiana Acad. Sci.* **51**, 13.

Weatherwax, P. (1954) *Indian Corn in Old America.* Macmillan, New York.

Weatherwax, P. (1955). Structure and development of reproductive organs. In *Corn and Corn Improvement.* p. 89. New York: Academic Press.

Wellhausen, E. J., Fuentes O., A. and Hernandez C., A. *in collaboration with* Mangelsdorf, P. C. (1957). Races of maize in Central America. *Nat. Acad. Sci. Nat. Res. Counc. Publ.* **511**, 1.

Wellhausen, E. J., Roberts, L. M. and Hernandez X., E., *in collaboration with* Mangelsdorf, P. C. (1952). Races of maize in Mexico. *Bussey Inst. Harvard Univ.*

Whyte, R. O. (1958). Plant exploration, collection and introduction. *F.A.O. Agric. Studies,* **41**, 1.

Williams, R. J. (1956). The significance of intraspecific variation in five Mediterranean perennial grass species. Thesis, University of Sydney.

Williams, W. (1945). Varieties and strains of red and white clover—British and foreign. *Welsh Pl. Breed. Sta. Bull.* **H16**, 1.

Woodworth, C. M., Leng, E. R. and Jugenheimer, R. W. (1952). Fifty generations of selection for protein and oil in corn. *Agron. J.* **44**, 60.

Yamashita, K., Tanaka, M. and Koyama, M. (1957). Studies on flour quality in *Triticum* and *Aegilops. Seiken Ziko,* **8**, 20.

Zohary, D. (1960). Studies on the origin of cultivated barley. *Bull. Res. Counc. Israel,* **9D**, 21.

INDEX

Åberg, 84
Abyssinia, 58
adaptation, 38, 120, 155, 163, 164
Aegilops, 72, 110, 112, 146, 152
  A. bicornis, 109, 118
  A. caudata, 106
  A. cylindrica, 106
  A. mutica, 110
  A. sharonensis, 109
  A. speltoides, 109, Pl. XV
  A. squarrosa, 78, 106, Pl. VII
A. umbellulata, 150
Afghanistan, 55
Africa, 57, 168, 169, 176
agricultural adaptation, 120
agriculture, beginnings, 166
Agropyron, 72, 83, 108
  A. repens, 48
  A. triticeum, 108
Agrostis tenuis, 159
allopolyploid, 71
Aitken, 156
aluminium toxicity, 160
Amaranthus, 167
amelioration of climate, 7
amphiploid wheat, 76
Anderson, 26, 35, 36, 41, 43, 99, 108
Andropogon, 153
aneuploid, 113
anthocyanin, 131
antiquity of crops, 167
añu, 132
apomixis, 165
arable cultivation, 22
Artemisia, 9
Argentine, 127
Argentine popcorn, Pl. II
Arrhenatherum tuberosum, 5
Australia, 51, 168,
Avdulov, 108
Avena, 72, 89, Pl. XI
  A. abyssinica, 92, Pl. XI
  A. barbata, 92, Pl. XI
  A. brevis, 92
  A. byzantina, 93, Pl. XI
  A. clauda, 90
  A. fatua, 93
  A. hirtula, 92, Pl. XI
  A. longiglumis, 90, Pl. XI
  A. ludoviciana, 94
  A. nuda, 92
  A. pilosa, 90, Pl. XI
  A. sativa, 93, Pl. XI
  A. sterilis, 94
  A. strigosa, 92, Pl. XI
  A. vaviloviana, 92
  A. ventricosa, 90
  A. wiestii, 92, Pl. XI
  see also oats
awn development, 72
Axonopus, 146, 147

Ball, 62
banana, 167, 169
Bandkeramic, 13
Bantu, 60
Barclay, 155
Barghoorn, 25, Pl. I
barley, 70, 81, 122, Pl. X
  agriocrithon, 82, 84, Pl. IX
  alleles, 86
  ancient, 4, 81
  botanical forms, 83
  Cerealia, 82, 83, 87
  chromosomes, 87
  deficiens, 87, 88
  disease resistance, 89
  distribution, 92
  heterospiculate, 81
  medicum, 88
  mutans, 88
  mutants, 87
  origin, 86
  pallidum, 88
  polyploidy, 82
  primitive, 84
  rachis, 86
  seed dispersal, 84, 88
  six row, 81, 87
  spikelets, 87
  spontaneum, 84, Pl. IX
  trypanocarpy, 85
  two row, 81, 87
  vulgare, 84
  weed, 85
  wild, 72, 82, 83, Pl. IX
Barrales, 160
basket-making, 50

Bat Cave, 26
Bateman, 135
Baxter, 134
bean, *see* legumes
Bear, 36
Beddgelert, 22
Beddows, 148
bees, 136
Bell, 80, 81, 88, 106, 168, 172
Bhatti, 66
Black, 163
Black Death, 22
Blakeway trackway, 15
Bleier, 106
blight-resistant cotton, 173
Boam, 67
bog burials, 2, 5
Bolivia, 127
Borrill, 146, 151, 155, 165
Bowden, 82, 84, 104
Bradshaw, 159
Brandwirtschaft, 9, 14
Brassica, 168
Brazil, 127
bread wheat, 104
Breckland, 11
breeding, 101, 165, 174
  for cultural conditions, 176
  methods, 99
Breggar, 36
brewing, 37, 50
Brieger, 42
Bromus, *see* grass
Brooke, 60
Brown, 40–2
Brücher, 133
Bukasov, 127, 130, 140
Burkhill, 61, 62

cacao, 169
caffeic acid glucoside, 131
Caistor, 20
Cameron, 42
Camelina, 166
carbohydrate, 167
Cárdenas, 131, 133
Carnahan, 153
cattle, 177
Celarier, 50, 51, 54
Centrosema, 147, 164
Ceratochloa, 152
cereals, *see* barley, maize, oats, rye, wheat
  age of, 100
  crosses, 96, 98
  weed forms, 100
Cervantes, 43
chalk downs, 20
Chalus, 107
Chapman, 98, 107, 110, 112, 115, 117, 119
Charles, 162
Chatterjee, 155
Chenopodium, 166
  album, 5
  quinoa, 132
chicha, 37
Chile, 123
Chinese Amber Canes, 62
chromosomes
  banana, 172
  homoeologous, 80
  maize, 28
  oats, 90, 92
  potato, 124
  rye, 98
  sorghum, 51, 54
  wheat, 112
chuño, 132
civilization and a wild grass, 122
Clark, 7, 13, 14
classification of cultivated plants, 71
Clausen, 165
Clayton, 66
clearance, 9, 11, 14, 17, 22
Clisby, 25
climate, 7, 151, 158, 170
clonal propagation, 172
clover
  Dollard, 160
  early flowering, 159
  Pilgrim, 163
  red, 148, 157, 159
  subterranean, 164
  white, 148, 159, 162
  see also forage, *Trifolium*
cocksfoot, *see* grass
coconut, 168
Coffman, 95
colchicine, 128
Collins, 35, 42, 38
Colombia, 38, 134
competition, response to freedom from, 46
Coombe, 153
corn, *see* maize, wheat, etc.
  flint, 37
  flour, 37

corn (*cont.*)
  sugary, 37
  sweet, 37
Cooper, 155–62, 172
Corn Belt dent corn, 29, 41
Corner, 131
Correll, 124, 131
Corylis, 16
cotton, 169
  disease-resistant, 171, 173
  for mechanical harvesting, 180
  G. herbaceum, 180
  G. hirsutum, 180
  introduction to Nile valley, 178
couch grass, 48
p-coumarylglucose, 131
Coxcatlán Cave, 26
crop evolution, 166, 178
cross pollination, 173
Cutler, 28, 42
cytogenetics, 103
cytotaxonomy, 153

Dactylis, 146, 151, *see* grass
Daday, 156, 159, 161
Danish bog, 24
Darwin, 140
dating, radiocarbon, 7, 13, 26
Davern, 160
day-length, 157
Desmodium, 147, 164
Dick, 26
diploid, *see* polyploid
diploidization, 119
disforestation, 22
disgenic practice, 39
distribution of crop
  barley, 92
  cotton, 169
  grass, 143
  legumes, 146
  limits of, 169
  maize, 26, 36, 37, 49
  Malaysian plants, 60
  oats, 95
  potato, 129
  rye, 98
  sorghum, 51–7, 61
  soy bean, 169
diversity, 172
Dobzhansky, 36
Dodds, 134–8, 141, 168, 174
Doggett, 54, 166, 168, 172, 179
domestication of maize, 29
dormancy, seed, 157
Drosophila, 67
  as domestic insect, 172
drought, 157, 158
  see also water
dung layers, 14
dura, 50

ecosystem, 7
Ecuador, 135
Edwards, 162
Elymus, 73
emmer wheat, 59
Endrizzi, 54
Ethiopia, 60
Euavena, 91
Eusorghum, 50, 51
Evans, A. M., 153
evolution, 166
evolution prospects, 178

farming systems, 142
Farquharson, 43
fenland, 20
Festuca, 153, 162
  see also grass
Fiedler, 133
field resistance, 177
Fisher, 172
flax, 59
flint corn, 37
flint mine, 11
flour corn, 37
flowering time, 158, 161
food plants, 142, 166
forage, 142–65
  cross-fertile, 160
  germination, 158
  lead resistance, 159
  leafy varieties, 164
  management, 162
  migration, 162
  taxonomy, 149
  testing, 149
  permanence, 163
Ford, 158
Frankel, 158, 163
Fransden, 133
Froghole, 22

Galinat, 27, 42, 43
Garber, 50, 51
genes, major and minor, 176
genetic
  barrier, 155

genetic (*cont.*)
characters, loss of, 102
diversity, 172
drift, 33, 37
manipulation, 70
variability, 40
genome, 74, 77, 174
Glastonbury lake village, 20
Godwin, 11–15, 166
Goldschmidt's 'hopeful monster', 30
Gottschalk, 133
germination, 159
glucoside, 131
grass, 142
Aegilops umbellulata, 150
Andropogon, 153
Bromus arizonicus, 153
B. carinatus, 153
B. catharticus, 153
B. curtipendula, 157
B. marginatus, 153
B. trinii, 152, 153
cold, effect of, 155
cocksfoot, 146, 148, 151, 156, 165
cross-fertile, 160
Dactylis, 146, 165
D. aschersoniana, 151
D. glomerata, 148, 151
D. lusitanica, 151, 165
D. santai, 151
D. smithii, 151
D. woronowii, 151
distribution, 143
early flowering, 158
Festuca, 153, 165
F. ovina, 159
F. pratensis, 162
forage, 142
frost resistant, 155
hybrids, 142, 151
Lolium, 150, 154
L. perenne, 148
L. rigidum, 149, 155, 156, 164
L. temulentum, 162
N. stricta, 158
origin, 146
potentialities, 142
Poa, 146, 153, 165
rainfall, effect 143
ryegrass, 148, 156, 158, 160
testing, 149
timothy, 163
variation, 143
wild ancestors, 142
Wimmera, 164
grain impressions in potsherds, 1
Gramineae, 71, 149, 167
grazing, 142
Grobman, 33, 41–5
groundnut, 168
Grun, 146
guinea-corn, 50
Gumban cultures, 59

Hackel, 50
Hadley, 56
Harborne, 131
Harlan, 100
Hartley, 143, 146
Haussdörfer, 128
Hawkes, 124, 130–4, 140, 141
Haynaldia, 96
hazel nuts, 16
Hector, 93
Helbaek, 1, 4, 5, 58, 74, 85, 100
hemp, 22
heterosis, 44
heterozygosity, 172
Hevea rubber, 166
Hill, 153
Hockham Mere, 111
Holcus, 50
Holme Fen, 19
homoeologous chromosomes, 80
partners, 112
hops, 22
Hordeum, 72, 82
Hougas, 138, 139
Howard, A., 173
Howard, G. L. C., 173
Howard, H. W., 124, 132, 133
husbandry, *see* management
Husk, maize, 24
Huskins, 114
Hutchinson, 67, 166
hybridization
cereals, 71, 96, 98
forage, 142, 165
grass, 151
maize, 38, 40
oats, 92
potato, 130, 133, 141
programme, 101
significance, 100
sorghum, 54, 56
wheat, 111

inbreeding, 103

India, 55, 169, 173
introgression, 100
Iran, 166
irrigation, *see* water
Italian rye-grass, *see* grass
Iversen, 9, 11

Jain, 97
Jarmo, 75, 85
jassid-resistant cotton, 171
Jenkins, 150
Jessen, 4
Jones, 94
Jordan, 166
jowar, 50
Jowett, 159
Juzepczuk, 140

Kafir corn, 50
Kamm, 86
karyotype analysis, 93
Keim, 38
Kempanna, 115
Kempton, 39, 42
Kidd, 66
Kiesselbach, 38
Kihara, 104–7, 146, 150
Kimber, 175
Knight, 176
Koyama, 120
Krongelb, 135
Kuleshov, 35
Kurz, 25

Lake village, 20
La Perra Cave, 32
Laude, 155, 163
Lead resistance, 159
Leakey, L. S. B., 59,
legumes, 142, 175
  cross-fertile, 160
  distribution, 146
  for protein, 168
  groundnut, 168
  Sainfoin, 168
  soy bean, 169
  Phaseolus, 147, 175
  see also clover, forage, lucerne
light intensity, 155, 160
Lilienfeld, 104
limits, geographical, 169
Linnaeus, 141
Lister, 42
Lolium, 150, 154, *see* grass
Long, 138
lucerne
  Canadian, 156, 165
  Hunter River, 165
  Ladak, 156, 165
  Ranger, 162
  Spanish Highland, 156
  winter growth, 159
  see also legumes
Ludwig, 160
Lupton, 88

macaroni wheat, 76
McFadden, 106, 108
McKee, 128
MacKey, 78, 104, 121
MacNeish, R. S., 26, 32, 166
Maguire, 43
maize, 23, 133, 166, Pl. I, III, IV
  allele, 28
  ancestors, 24
  archaeological sites, 26
  domestication, 46
  environmental adaptation, 38
  evolution, 29
  fossil 25, Pl. I
  genotypes, 28
  glume, 23, 32
  husk, 24, 32
  hybridization, 38, 40
  low position of ear, 47
  mesocotyl, 38
  mutability, 43
  origin, 25
  Peruvian flour corn, Pl. II
  prehistoric, Pl. I
  radiocarbon dating, 26
  resistance to drought and heat, 43
  resistance to insects and rust, 38
  rogue, 28
  seed dispersal, 24, 31, 32
  selection, 40
  sheathed head, 180
  silk, 23
  spikelet, 23, 33
  stunt disease, 43
  suckers, 47
  tillers, 47
  waxy, 35
  wild, 32
Malheiras-Gardé, 124
Malaysian plants in Africa, 60
Malzew, 91, 95
management, 142, 162, 170, 180

manganese toxicity, 160
Mangelsdorf, 23, 26–30, 36, 39, 42–4, 168
Marchioni, 43
Marks, 128, 129
Mather, 160
Meare lake village, 20
mechanized cultivation, breeding for 177, 180
Medicago, 147, 148
  see also legumes, lucerne
meiotic conjugation, 114, 117
meiotic behaviour, 54, 55
Mexico, 26, 124, Pl. I
Middle East culture, 12
migration, crop, 142, 162, 164, 178
mildew, powdery, 88, 99
millet, 50, 60, 81
milo, 50
Moench, 50
Morley, 150, 153, 156–9, 163, 165
Morrison, 17, 94
mutation, 43, 100, 118
Murdock, 58

Nardus stricta, 158
Nash, 50
Neobromus, 152
Neolithic agriculture, 59
Nile valley, 178
nitrophilous plants, 8
Njoro River Cave, 59
nobilization in sugar cane, 176
North America, 178

oak forest, 1
oats, 70, 89, Pl. XI
  adventive, 89
  ancient, 4, 166
  association of forms, 90
  awns, 90
  B. genome, 92
  chromosomes, 90, 92
  distribution, 95
  Euavena, 91
  as forage, 92
  genomes, 92, 94
  disease resistance, 95
  hybridization, 92
  hexaploid, 96
  karyotype analysis, 93
  panicle morphology, 94
  polyploidy, 90, Pl. XI
  spikelets, 90
  shedding, 90
  weed form, 89, 93, 96
  wild, 72, 89, 93, 94
oca, 132
Ochoa, 137
oil palm, 168
oil, vegetable, 168
Oinuma, 83, 87
Okamoto, 113
Oliver, 26
Olmsted, 157
Onobrychis, 147
organic sediments, 6
origins of cultivated plants, 1
Oxalis tuberosa, 132

pachytene stage of meiosis, 133
Pakistan, 55
palm, oil, 168
Pandey, 124
Panicum, 146, 147
Paraguay, 127
parthenocarpy, 174
Paspalum, 146, 147
pasture, *see* forage
Paxman, 134–7
Peak, 156, 159, 165
Peloquin, 138, 139
Perlova, 132
Peru, 27
Peruvian flour corn, Pl. II
Peters, 133
Peterson, 160
pH, low 160
Phalaris tuberosa, 149, 155, 164
Phaseolus, 147, 175, *see* legumes
phenotypic segregation, 134
philology, 103
Phleum pratense, 162
Phytophthora infestans, 126
pigment variability, 131
plant remains
  ancient, 20
  fossilized, Pl. I
  in bog-burials, 2
Plantago lanceolata, 9
plant breeding, 99, 101, 174
Pleistocene, 151
plough, 20
  scratch, 17
Poa, 146, 153, 165
pod corn, 27, Pl. II
  crossed with popcorn, 31
  fertile, 31

pod corn (*cont.*)
relict, 30
sterile, 29
pod-popcorn, 31, Pl. II
pollen
analysis, 5, 18
Corylis, 16
dating, 12, 25
diagrams, 6, 21
fossil, Pl. I
frequency, 6
grass, 9
Hockham, 11
pollination
by bees, 136
cross, 173
wind, 172
Polygonum, 5, 166
polyploid, 90, Pl. XI
barley, 82
cereals, 71
crop plants, 175
Gramineae, 175
grass, 146, 149
Leguminosae, 175
mechanism, 119
potato, 125, 129, 139
sorghum, 51, 55, 56
wheat, 30, 74, 103, 104, 120
popcorn, 27
crossed with pod corn, 31
Popenoe, 42
post-glacial changes, 151
potato, 123, 166
alkaloids, 131
alleles, 134
ancestors, 128
berries, 126, 138
British Isles, in, 140
clones, 135, 136
cultivars, 131
diploid, 129, 136
distribution, 129
domestication, 133
early, 129
Europe, in, 141
fertility, 134
flavour, 129, 136
flowers, 125, 126, 139
fossil, 123
haulm, 131
hybrids, 130
incompatibility, 124
intercrossing, 138
leaf, 126, 131, 137
leaf-index, 140
meiosis, 128
morphology, 131
non-dormant, 137
ploidy, 129
rosette, 130
short day, 137
soups and stews, in, 136
taxonomy, 124, 133
tubers, 131
variability, 129
yellow, 136
weed, as, 128
wild, 123, 128, 130, 132
potsherds, grain impressions in, 1
powdery mildew, 88, 96
prehistoric agriculture, 1
Price, 97
protein, 168

querns, 12
quinoa, 132
rachis, tough, 177
radiocarbon dating, 7, 13, 26
Rajhathy, 94
Randolph, 42
Rao, 78
Raper, 135
Reeves, 26, 27, 30
Reid's Yellow Dent, 40
relict character, 30
Rick, 173
Riley, 76, 79–81, 98, 106–19, 129, 175, 176
Rivet wheat, 76
Roberts, 33
Rockley Down, 5
rogue maize, 28
root system, nourishment of, 48
Ross, 133, 138
rubber, 166
Rupper, 42
rust, 38, 99
rye, 70, 96, Pl. XII
Agrestis, 97
ancient, 97, 166
Cerealia, 97
chromosomes, 98
compatibility, 97
disease resistance, 99
distribution, 98
drought resistance, 98
hairy neck, 97

rye (*cont.*)
  hybrids, 97
  hybrids with other cereals, 96, 98
  investment of grain, 98
  perennial, 97
  poverty crop, as, 98
  weed, as, 97
  wild, 72, 97
  winter-hardiness, 98
ryegrass, *see* grass

Sainfoin, 168
  see also legumes
Saint-Hilaire, 27
Sakamura, 104
Salaman, 140, 141
saltation, 48
Sarkar, 76, 108
Sax, 106
Schick, 128
Schiemann, 104
Schizophyllum, 134
Schreiber, 131
Schreiter, 133
Schulze, 157
scratch plough, 17
Sears, E. G., 113, 114
Sears, E. R., 80, 106–9
Sears, P., 25
Secale, 4, 76, 96
  S. africanum, 97
  S. cereale, 97, Pl. XII
  S. montanum, 97
seed crop, 121
seed dispersal, 24, 32, 84
segregation, phenotypic, 134
selection pressure, 163, 178
self-fertilization, 174
Senn, 153
sexual reproduction, loss of, 174
Shebeski, 80
Shippea Hill, 12
shucks, 38
Sielinger, 68
silk (maize), 23
Simmonds, 130, 140
Sitanion, 83
Smith, C. E., 26, 39, 42
Smith, D., 162
Smith, L., 88
Snaydon, 159
Snowden, 50, 51, 56–8, 64
soil, effect of, 179
Solanum, 123, *see also* potato
Solanum × *vallis-mexici*, 128
Solanum acaule, 127, 130
S. agrimonifolium, 126
S. bulbocastanum, 124, 131
S. canasense, 132
S. capsicibaccatum, 127
S. chacoense, 127
S. Commersonii, 127
S. demissum, 126, 132
S. guerreroense, 128
S. herrerae, 132
S. infundibuliforme, 127
S. leptophyes, 132
S. leptostigma, 132
S. longipedicellatum, 132
S. molinae, 132
S. morreliforme, 125
S. oxycarpum, 126
S. polyadenium, 124, 126
S. reconditum, 126
S. soukupii, 132
S. sparsipilum, 132
S. stoloniferum, 126
S. stenotomum, 132
S. vernei, 133
S. verrucosum, 128
S. Wittmackii, 126
S. Woodsonii, 126
S. zerophyllum, 127
Sorghastrum, 50, 51
sorghum, 50, 166
  awnless, 63
  chromosomes, 51, 54
  classification, 51, 66
  clines, 66
  distribution, 52–7, 61
  diversification, 67
  duality of species, 67
  Ethiopia, in, 60
  genome, 56
  groups, 51, 65
  hybridization, 50, 62, 68
  infertile, 66
  management, 170
  nomenclature, 66
  origin, 58
  polyploids, 51, 68
  rainfall, effect of, 65
  rhizomes, 54
  sheathed head, 179
  species number, 65
  sterility, 54, 64
  types, 51, 62, 65, 170
  variability, 63

sorghum (*cont.*)
  wild, 62
Sorghum
  S. caffrorum, 66
  S. arundinaceum, 65
  S. subglabrescens, 66
  S. virgatum, 66
  S. vulgare, 66
  S. bicolor, 66
South America, 122, 126, 131
spikelet
  barley, 87
  cereals, 72
  maize, 23, Pl. III, IV
  oats, 90
  sorghum, 50, 64, 67
  wheat, 58, 108
Sprague, 35
Stanford, 163
Stapf, 50, 56, 57
Stebbins, 71–3, 76, 108, 146–52
Steppler, 160
sterility
  banana, 174
  potato, 124
  root plants, 174
  sorghum, 54, 55
  wheat, 111
Stonor, 26, 35, 36
stunt disease, 43
Sturtevant, 27
suckers, maize, 47
Sudan grass, 67
sugar cane, 175, 176
sunflower, 168
survival, 32, 174
Swaminathan, 78, 124
sweet corn, 37
sweet potato, 167
Sylven, 162
synapsis, 116

Takahashi, 85, 88
Tanaka, 107, 120
Tanganyika, 179
taxonomy, 71, 149
teosinte, 168, Pl. III, IV
Thoday, 67
Thomas, 165
tillers, 47
Tollund Man, 4
tomato, 173
Topeth, 59
toxicity
  aluminium, 160
  lead, 159
  manganese, 160
trackway of wood, 15, 154
Trifolium, 147, 148, 154
  T. occidentale, 153
  T. pratense, 148, 162
  T. repens, 148, 153
  T. subterraneum, 149, 153, 156, 158
  see also clover, legumes
Trigonella, 147
tripsacoid cobs, 42
Tripsacum, 33, 43
Triticeae, 72, 73
Triticum, 11, 72, 75, 103, *see also* wheat
  T. aegilopoides, 76, 109, Pl. V
  T. aestivum, 79, 103, Pl. VIII, XV
  T. armeniacum, 75
  T. compactum, 75, 78, Pl. VIII
  T. carthlicum, 77
  T. dicoccoides, 75, 77, Pl. VI, VII
  T. dicoccum, 75, 109, Pl. VI
  T. durum, 75, 77, 109
  T. longissima, 109
  T. macha, 75, 78, Pl. VIII
  T. monococcum, 75, 76, 109, Pl. XV
  T. orientale, 75, 77
  T. persicum, 75, 77
  T. polonicum, 75
  T. pyramidale, 75
  T. rigidum, 75
  T. spelta, 75, 76, 106, Pl. VII, XV
  T. speltiforme, 75
  T. sphaerococcum, 75, 78, Pl. VIII
  T. thaoudar, 75, 106
  T. timopheevi, 75, 77, 104
  T. tubalicum, 75
  T. turgidum, 75, 77
  T. vavilovi, 75, 78, Pl. VIII
  T. vulgare, 75
  T. zhukovskyi, 78
Troels-Smith, 11, 13, 14
Tropaeolum tuberosum, 132
Tunis grass, 66
Turkey, 166
Turner, 16, 17

ulluco, 132
Ullucus tuberosum, 132
umbelliferone, 131
Unrau, 110, 119
Uruguay, 127
U.S.A., 27, 123

variability, 99, 100, 172
Vavilov, 58, 59, 99
Vavilov distribution, 173
vegetable oil, 167
Vose, 160

Wagenaar, 105
Walker, 22
Wallace, 40, 41
Wangenheim, 133
Watkins, 77
water supply, 43, 98, 143, 154, 157, 164
waxy maize, 35
Weatherwax, 23, 39
Wellhausen, 33, 41, 43
weed
  barley, 85
  cereal, 100
  oats, 89, 93, 96
wheat, 4, 70, 73, 89, 103, 122, 150
  allele, 119
  amphiploid, 76
  awn, 109
  bread, 104
  chromosomes, 78, 104, 113, 116, Pl. XIII, XIV, XVI
  chromosome V, 116, 121
  diploid, Pl. V
  emmer, 59
  fertile, 120, 121
  free-threshing, 76
  genome, 74, 104
  glume, 77, 109
  haploid, 105
  hexaploid, 76, 104, Pl. VII, VIII, XV
  hybrids, 104
  hybrid with rye, 98
  karyotype, 77, 108
  K. factor, 78
  mutation, 118, 120
  Neolithic, 4
  Old World, 166
  origin, 110
  parthenogenetic, 105
  polyploid, 30, 74, 75, 104
  primitive, 58
  Q. factor, 78
  Rivet, 76
  seed, 121
  sterile, 105, 121
  tetraploid, Pl. VII
  wild, 72, 76, 110
Whixall Moss, 17
Whyte, 146, 147, 164
wild ancestors, 167
wild grasses, 56
Williams, 148, 157
Willis, 12
Wimmera ryegrass, 164
wooden trackway, 15
woodland clearance, 9, 11, 14, 17, 22

yam, 167
Yamashita, 120

Zea mays, 33
  see also maize
Zea mexicana, Pl. III
Zohary, 84, 146, 151